ULTIMATE SLIME

EXTREME EDITION

100 New
Recipes & Projects
for Oddly Satisfying,
Borax-Free Slime

ALYSSA JAGAN

QUARRY

Brimming with creative inspiration, how-to projects, and useful information to enrich your everyday life, Quarto Knows is a favorite destination for those pursuing their interests and passions. Visit our site and dig deeper with our books into your area of interest: Quarto Creates, Quarto Cooks, Quarto Homes, Quarto Lives, Quarto Drives, Quarto Explores, Quarto Gifts, or Quarto Kids.

Inspiring | Educating | Creating | Entertaining

First Published in 2020 by Quarry Books, an imprint of The Quarto Group,
100 Cummings Center, Suite 265-D, Beverly, MA 01915, USA.
T (978) 282-9590 F (978) 283-2742 QuartoKnows.com

Quarry Books titles are also available at discount for retail, wholesale, promotional, and bulk purchase. For details, contact the Special Sales Manager by email at specialsales@quarto.com or by mail at The Quarto Group, Attn: Special Sales Manager, 100 Cummings Center, Suite 265-D, Beverly, MA 01915, USA.

10 9 8 7 6 5 4 3 2 1

ISBN: 978-1-63159-827-2

Digital edition published in 2020
eISBN: 978-1-63159-828-9

Library of Congress Cataloging-in-Publication Data is available

Page Layout: Megan Jones Design
Photograpy: Sam Welbourn Photography, www.samwelbournphotography.com;
except where indicated

Printed in China

CONTENTS

Introduction **6**

Making Slime Safely 7

1

SLIME STARTUP 9

Essential Slime Supplies...................... **10**

Basic White Glue Slime...................... **12**

Extra-Thick White Glue Slime **14**

Basic Clear Glue Slime...................... **16**

**Make It Right: Fixes for
Sticky Situations** **18**

Cleaning Up and Storing Slime **20**

2

ADD-INS FOR EPIC SLIME 23

Must-Have Mix-ins............................... **24**

Colorants 24

Pigments 24

Fragrance Oils 25

Glitter 25

Foam Beads 26

Plastic Beads: Fishbowl,
Sugar, and Slushie 26

Next-Level Mix-ins........................... **27**

Pompoms 27

Acrylic Charms 27

Metallic Foil 28

Make Your Own Mix-ins.................... **29**

Fake Cereal................................. 29

Fake Sprinkles and Chocolate Chunks.......30

Charms 32

Foam Cubes................................. 33

Foam Shapes 34

Foam Chocolate Chunks.................35

3

SECRETS OF CLAY AND
SNOW SLIME REVEALED!.............37

**Clay Play: Adding Air-Dry Clay
to Slime** **38**

Inflating Slime.............................40

Bread Slime 42

**Let It Snow: Slime and Expandable
Fake Snow** **44**

Jelly Slime 46

Icee Slime 48

Cloud Creme 50

Cloud Slime 52

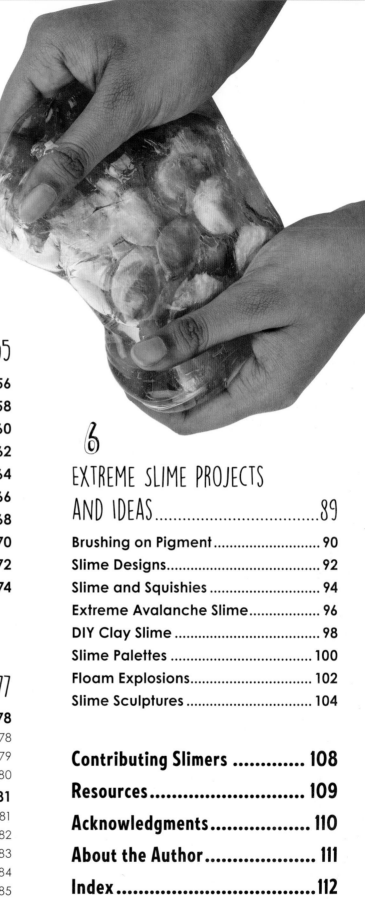

4

EXTREME SLIME RECIPES 55

Color-Changing Slime 56
Pompom Slime 58
Chocolate Chunk Slime 60
Cereal Slime 62
Foam Shapes Slime 64
Crunchy Charms Slime 66
Jelly Cube Slime 68
Metallic Foil Slime 70
Fizz Slime 72
Kawaii Slime 74

5

HYBRID SLIMES 77

Crunchy Hybrid Slimes 78
  Charm Slime + Fishbowl Slime 78
  Fizz Slime + Slushie Slime 79
  Fizz Slime + Inflating Slime 80
Fluffy Hybrid Slimes 81
  Cloud Slime + Clear Slime 81
  Jelly Slime + Clear Glue Floam 82
  Inflating Slime + Clear Glue Floam 83
  Jelly Cube Slime + Glossy Slime 84
  Inflating Slime + Icee Slime 85
All the Feels Hybrid Slimes 86

6

EXTREME SLIME PROJECTS AND IDEAS 89

Brushing on Pigment 90
Slime Designs 92
Slime and Squishies 94
Extreme Avalanche Slime 96
DIY Clay Slime 98
Slime Palettes 100
Floam Explosions 102
Slime Sculptures 104

Contributing Slimers 108
Resources 109
Acknowledgments 110
About the Author 111
Index 112

INTRODUCTION

Slime has changed my life. Ever since this ooey, gooey substance came into my life, I've been hooked.

Slime has so many different purposes. Firstly, slime is a stress reliever. I play with it while doing my math homework or when I need to destress after a long day.

Slime is also fun. There are so many interesting ways to make and play with slime! It's such a great craft—it's the perfect activity to do alone, but it's also a fun way to spend time with your friends. It's great to knead the more dough-like slimes and to poke thick slimes.

And, of course, slime is a creative outlet. You can customize every single aspect of slime and make it your own. You can also make slime as simple or as intricate as you want. There are endless possibilities with slime, and there are still new slimes being created and discovered.

After writing my first slime recipe book, *Ultimate Slime*, I knew I had more to share with you. In this book, there are more than 100 all-new slime recipes and projects that are better than ever and take slime to the next level—that's why it's the *Extreme Edition*!

Back in 2016, slime took social media by storm, and the community is still thriving. I began posting about my slimey creations in the summer of 2016 and have since gained a following of more than 800,000 on my Instagram account, @craftyslimecreator. I also share longer videos on my YouTube channel, Alyssa J.

Slime is truly an amazing hobby, and if you follow the steps in this book, it's fairly straightforward. You can create the master slimes in many different ways. An important part of slime making is activating the slime, which means that the slime forms a blob and creates the slimey texture. It's your and your parents' decision to choose the activator you're going to use. Make sure you do your research and discuss this with your parents before making slime. Keep in mind that regardless of which activator you choose, a chemical reaction will take place. Also be sure to check that you don't have any allergies to any ingredients that you use to make slime.

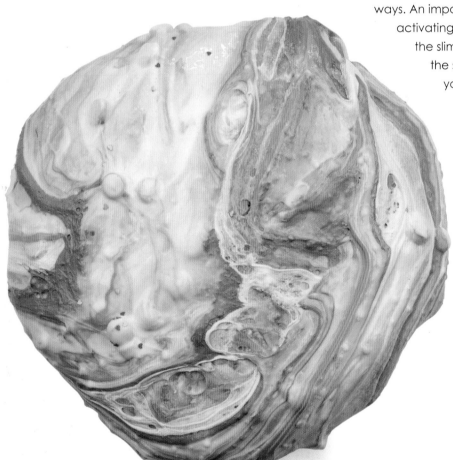

Please keep in mind that in this book I'm sharing examples of what I—and some lovely contributors—have done with slime, and that you can experiment to your heart's content! Experiment with ratios of ingredients, colors, and textures! Remember to have fun with slime.

Don't forget to post videos and photos of your creations on Instagram and to use the hashtag #craftyslimecreations so I can see the slimes you make. The slime community on social media is so welcoming. Can't wait to see you there!

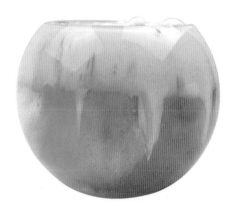

MAKING SLIME SAFELY: A NOTE FOR ADULTS

- **Always supervise children closely** when they're working with slime or any of its ingredients. Handle all products used to make slime carefully.

- **Do not use a product if you or a child has a sensitivity to it,** and discontinue the use of any product if you or a child has an adverse reaction to it.

- **Never eat slime,** and *never* let children put slime or any of its ingredients in their mouths.

- **Always label slime so it won't be mistaken for food.** It's especially important to label food replicas that are made with slime to avoid having someone mistake them for food.

- **Always store slime out of the reach of children,** especially young children and pets.

- **Do not prepare slime on surfaces or in areas used to prepare or serve food.**

- **Always use disposable bowls and utensils to make slime.** Or set aside a set of bowls and utensils that you use only for slime making, and *never* reuse them for food preparation or for bathing.

- **Avoid getting slime on skin or clothing.**

- **Always have adequate ventilation** when making slime.

- **Always make and store slime at room temperature.** Never freeze or heat slime.

- **When throwing slime away, make sure you're in compliance with local, state, and federal laws** for disposing its ingredients.

- **After handling slime, always wash hands thoroughly** with soap and water.

1 Slime Startup

THIS CHAPTER IS THE BASIS FOR THE REST OF THE RECIPES IN THIS BOOK. IF YOU MASTER THESE BASE RECIPES AND LEARN WHAT ALL THESE INGREDIENTS DO, THEN YOU'LL BE A SLIME PRO! THE MORE YOU EXPERIMENT WITH THESE INGREDIENTS, THE MORE FUN SLIME MAKING WILL BE.

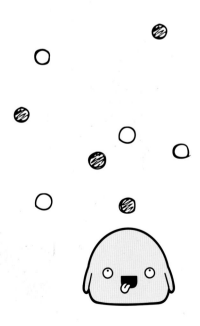

ESSENTIAL SLIME SUPPLIES

In this section, you'll find all the necessary ingredients for slime. Slime consists of two main components: a glue base and activator. You can add other ingredients to change the texture.

Remember to have an adult with you to help so that they can make sure you use the ingredients safely.

POLYVINYL ACETATE (PVA) GLUE

PVA glue is essential to slime making. **Make sure that you use only nontoxic glues.** To test if a glue contains PVA, put some on paper and let it dry. If it's clear when it's dry, then it contains PVA. The recipes in this book use two types of glue: white glue and clear glue.

I use two **white glues**: Elmer's School Glue and Elmer's Glue-All. These glues have very different consistencies. School Glue has a thinner consistency, which creates stretchier, glossier slimes. Glue-All has a much thicker, heavier consistency, resulting in slimes that are glossy when you let them sit, but become matte when you play with them.

Depending on the activator, there are different ways to make slime with these glues. I recommend starting with School Glue, as it produces the best product consistently, whereas Glue-All is a more advanced glue. For further details, see the Basic White Glue Slime recipe on page 12.

For **clear glue,** I use Elmer's Clear Glue, specifically in the quart (946 ml) size. Currently, Elmer's Clear Glue in the quart size (946 ml) produces the clearest slime; the glue in the gallon (3.8 L) size can have a yellow tint. You can use the gallon (3.8 L) size if you're planning to color your slime since the yellow won't show, but for perfectly clear slime, use the quart (946 ml) size.

Each recipe requires a different glue and specifies which one you'll need. For example, Cloud Slime (see page 52) must be made with white glue, but Icee Slime (page 48) must be made with clear glue to create the correct, sizzly texture.

ACTIVATORS

An activator is crucial to making slime. When you add the perfect amount, it activates the glue's polymer (plastic-like) molecules so they become stretchy and create the signature slimey texture. I use three borax-free options.

Baking soda and contact lens solution. You must use these two products together to activate slime. The contact lens solution you use must have boric acid or buffered saline listed in the ingredients in order for it to make slime. I've found that Renu Fresh Multipurpose Solution consistently makes the best slime. Keep in mind that this activator is the only borax-free option that keeps clear slime clear; the others make it foggy. An issue I've experienced with this slime is that it doesn't last as long as slimes made with other activators. I recommend using liquid starch instead. if that is not available, or if you want to make clear, clear glue–based slime, use contact lens solution and baking soda and follow the tips on pages 18–21. As shown in the slime base recipes (pages 12 and 16), I recommend adding glycerin to slimes activated using contact lens solution and baking soda to prevent the slime from developing a hard layer on top.

Liquid laundry detergent. This must contain boric acid or a borate ion, such as Tide Free and Gentle, the brand I use.

I've found that this activator doesn't create the hard layer on top that the baking soda and contact lens solution does, but it has a slight odor to it. This activator also makes clear slime cloudy.

Liquid starch. My recommended brand is Sta-Flo because I consistently make gorgeous stretchy slime with it. This activator also makes clear slime cloudy, but of all three activators, this is my personal favorite.

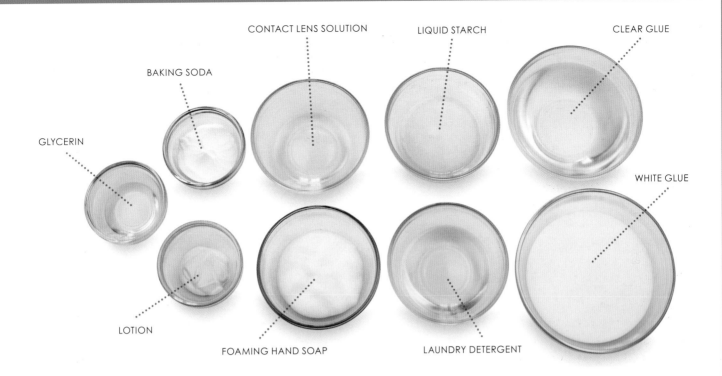

GLYCERIN · BAKING SODA · CONTACT LENS SOLUTION · LIQUID STARCH · CLEAR GLUE · WHITE GLUE · LOTION · FOAMING HAND SOAP · LAUNDRY DETERGENT

OPTIONAL ADD-INS

These add-ins change the texture of your base slime. It's exciting to experiment with different products to see the various outcomes. When you add in these products, they'll affect the slime's color and scent. For example, if you add pink, floral-scented foaming hand soap, your slime will turn slightly pink and have a floral scent. Also, if you're adding several different products, make sure that they're unscented, or that their scents match. See page 25 for more details about fragrance oils, which you can use to add scent to slime.

Foaming hand soap. I use this in almost all my slimes. Alone, it doesn't activate slime, but during the activation process, I find that it helps the slime to form and clump together. Overall, it makes the slime stretchier and last longer. Make sure the hand soap is the foaming type and has the proper foaming dispenser or it won't work. When adding foaming hand soap to clear slime, make sure the liquid is colorless and clear. I like using it in clear slimes, but I find that it does affect the clarity, so be aware of that when making clear slime.

Glycerin. Glycerin acts as a "deactivator," in the sense that it will undo the activation process. I mainly add it to slimes activated with contact lens solution and baking soda to counteract the hard layer that tends to form on top. (See the tip on page 18 for more details.) Keep in mind that adding too much will liquefy your slime, so make sure you add it in small amounts. Glycerin also reduces the stickiness of slime when you use the right amount, so it may cause small elements to fall out. I recommend experimenting with glycerin to see how it reacts with your slime and the specific products/brands you're using.

Lotion. Lotion will make your slime much stretchier, but will make clear slime cloudy. Any type of lotion works, but its scent and color will affect your slime.

BASIC WHITE GLUE SLIME

THIS IS THE MOST BASIC SLIME, AND IT'S VERY IMPORTANT TO LEARN FIRST. ONCE YOU MASTER THIS AND THE BASIC CLEAR GLUE SLIME ON PAGE 16, YOU'LL BE A PRO! THIS RECIPE MAKES A GORGEOUS GLOSSY SLIME THAT MAKES NICE CLICKY SOUNDS. THIS BASE SLIME IS USED IN MANY OTHER SLIMES IN THIS BOOK.

Banana Bonanza Slime

WHAT YOU'LL NEED

Equipment

Large bowl
Measuring cups and spoons
Mixing tool (spoon, spatula, or stir stick)
Airtight container

Ingredients

1 cup (235 ml) white PVA glue (Elmer's School Glue)
Approximately ½ cup (120 ml) foaming hand soap
An activator (see chart on page 13 for details)

Optional

Color additive (see page 24)
1 to 3 drops of fragrance oil

1. Place the glue in a large bowl. Add foaming hand soap to the glue (A). If you want, add color and/or fragrance oil (see pages 24 and 25 for some options). Mix thoroughly. To make the sample slime, I used yellow food coloring.

2. Add your activator to the glue in small amounts, no more than 2 tablespoons (28 ml) at a time (B). See the guidelines on page 13 for the recommended amounts of each activator. Remember to add only one type of activator to each batch of slime.

BASIC WHITE GLUE SLIME RECIPES

- Inflating Slime (page 40)
- Bread Slime (page 42)
- Cereal Slime (page 62)
- Chocolate Chunk Slime
- Foam Shapes Slime (page 64)
- Hybrid Slime: Cloud + Clear (page 81)
- Hybrid Slime: Fizz + Inflating (page 80)
- Hybrid Slime: Inflating + Clear Glue Floam (page 83)
- Hybrid Slime: Inflating + Icee (page 86)
- Hybrid Slime: Jelly Cube + Glossy (page 84)
- Kawaii Slime (page 74)

BASIC WHITE GLUE SLIME PROJECTS

- DIY Clay Slime (page 98)
- Extreme Avalanche Slime (page 96)
- Slime Designs (page 92)
- Slime Palettes (page 100)
- Slime Sculptures (page 104)

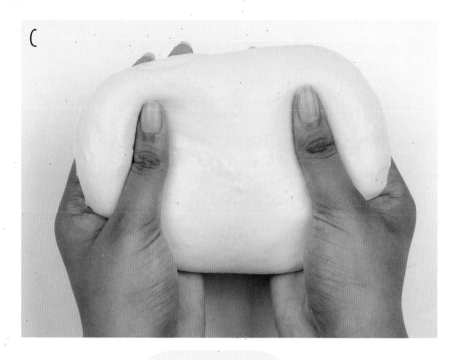

 Once the mixture is slightly sticky, start to knead the slime (C). If the mixture is too sticky, dip your fingers in a little activator before kneading so less slime will stick to your hands. Playing with slime is the best way to mix it fully and get the best possible texture.

Store the finished slime in an airtight container so it doesn't dry out. For the best glossy texture, let the slime sit for 2 to 4 days. The slime glosses up because the air incorporated into the slime rises up and out over the 2 to 4 days.

Yield: Approximately 14 ounces (410 g)

ACTIVATOR	RECOMMENDED AMOUNT
Liquid laundry detergent	5 to 7 tablespoons (75 to 105 ml)
Liquid starch (such as Sta-Flo)	6 to 8 tablespoons (90 to 120 ml)
Baking soda + contact lens solution	Add 1 tablespoon (15 ml) of baking soda to the glue mixture and then add 1 to 3 tablespoons (15 to 45 ml) of contact lens solution. Renu Fresh Multipurpose Solution works well; if you use other brands, you may need to add a lot more. I also recommend adding 1 teaspoon (5 ml) of glycerin to the glue mixture if you use this activator, although this is optional. Slime made with this activator tends to develop a hard layer on top after a week, so to combat this, I recommend adding glycerin, which keeps it stretchy longer.

BASIC WHITE GLUE SLIME: HOW MUCH ACTIVATOR DO I NEED?

It depends on which activator you choose! See the recommended amounts at left.

Add no more than 2 tablespoons (28 ml) of activator at a time and then stir well. Check the consistency of the slime before adding more. The amount of activator you add will affect the consistency; you can adjust it to a consistency you like or one you want for a specific recipe.

If you add too much activator, your slime will become hard, but if you don't add enough, it will be too sticky.

If you make a mistake and need to fix your slime, go to the troubleshooting section on page 18–21.

Remember—use only one type of activator for each batch of slime!

EXTRA-THICK WHITE GLUE SLIME

THE SECRET TO MAKING EXTRA—THICK WHITE GLUE SLIME IS TO USE A DIFFERENT TYPE OF GLUE——ELMER'S GLUE-ALL! THIS GLUE REQUIRES A DIFFERENT AMOUNT OF ACTIVATOR THAN ELMER'S SCHOOL GLUE, SO MAKE SURE YOU USE THE RIGHT AMOUNT OF ACTIVATOR WITH EACH SPECIFIC GLUE.

Marshmallow Slime

WHAT YOU'LL NEED

Equipment

Large bowl
Measuring cups and spoons
Mixing tool (spoon, spatula, or stir stick)
Airtight container

Ingredients

½ cup (120 ml) white PVA glue (Elmer's Glue-All)
½ cup (120 ml) white PVA glue (Elmer's School Glue)
Approximately ½ cup (120 ml) foaming hand soap
An activator (see chart on page 15 for details)

Optional

Color additive (see page 24)
1 to 3 drops of fragrance oil

Yield: Approximately 14 ounces (410 g)

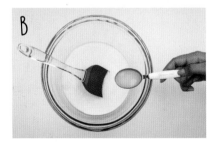

1. Place the glue in a large bowl. Add foaming hand soap to the glue **(A)**. If you want, add color and/or fragrance oil (see pages 24 and 25 for some options). Mix thoroughly.

2. Add your activator to the glue in small amounts, no more than 2 tablespoons (28 ml) at a time **(B)**. See the guidelines on page 15 for the recommended amounts of each type of activator—remember to add only one type of activator to each batch of slime.

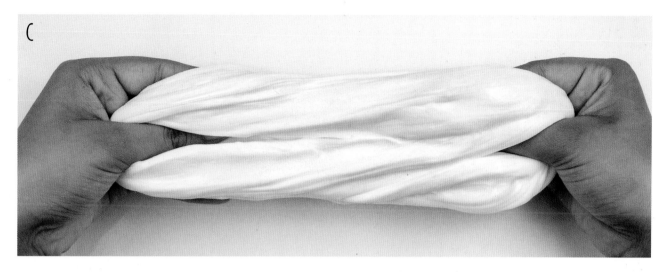

 3 Once the mixture is slightly sticky, start to knead the slime (C). If the mixture is too sticky, dip your fingers in a little activator before kneading so less slime will stick to your hands. Playing with slime is the best way to mix it fully and get the best possible texture.

4 Store the finished slime in an airtight container so it doesn't dry out. For the best glossy texture, let the slime sit for 2 to 4 days. The slime glosses up because the air incorporated into the slime rises up and out over the 2 to 4 days.

ACTIVATOR	RECOMMENDED AMOUNT
Liquid laundry detergent	4 to 6 tablespoons (60 to 90 ml)
Liquid starch (such as Sta-Flo)	4 to 6 tablespoons (60 to 90 ml)
Baking soda + contact lens solution	Add 1 teaspoon (5 ml) of baking soda to the glue mixture and then add 1 to 3 tablespoons (15 to 45 ml) of contact lens solution. Renu Fresh Multipurpose Solution works well; if you use other brands, you may need to add a lot more. I also recommend adding 1 teaspoon (5 ml) of glycerin to the glue mixture if you use this activator, although this is optional. Slime made with this activator tends to develop a hard layer on top after a week, so to combat this, I recommend adding glycerin, which keeps it stretchy longer.

EXTRA-THICK WHITE GLUE SLIME: HOW MUCH ACTIVATOR DO I NEED?

It depends on which activator you choose! See the recommended amounts at left.

Add no more than 2 tablespoons (28 ml) of activator at a time and then stir well. Check the consistency of the slime before adding more. The amount of activator you add will affect the consistency; you can adjust it to a consistency you like or one you want for a specific recipe.

If you add too much activator, your slime will become hard, but if you don't add enough, it will be too sticky.

If you make a mistake and need to fix your slime, go to the troubleshooting section on page 18–21.

Remember—use only one type of activator for each batch of slime!

BASIC CLEAR GLUE SLIME

THIS RECIPE FOR BASIC CLEAR GLUE SLIME IS AN ESSENTIAL RECIPE THAT'S USED IN LOTS OF OTHERS IN THIS BOOK, SO FIGURING OUT YOUR FAVORITE WAY TO MAKE IT IS IMPORTANT.

JUST AS WITH BASIC WHITE GLUE SLIME (PAGE 12), IF YOU ADD TOO MUCH ACTIVATOR, CLEAR GLUE SLIME WILL BECOME HARD, BUT IF YOU DON'T ADD ENOUGH, IT'LL BE TOO STICKY. IF YOU MAKE THESE MISTAKES AND NEED TO FIX YOUR SLIME, CHECK THE TROUBLESHOOTING SECTION ON PAGES 18–21.

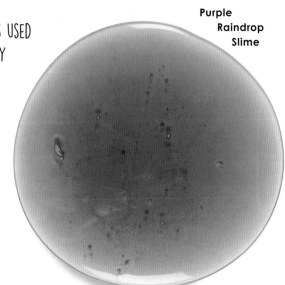

Purple
Raindrop
Slime

WHAT YOU'LL NEED

Equipment

Large bowl
Measuring cups and spoons
Mixing tool (spoon, spatula,
    or stir stick)
Airtight container

Ingredients

1 cup (235 ml) clear PVA glue
    (such as Elmer's Clear Glue)
An activator (see chart page 16
    for details)

Optional

Color additive (see page 24)
1 to 3 drops of fragrance oil (Please
    note that this may make clear
    glue–based slimes cloudy.)

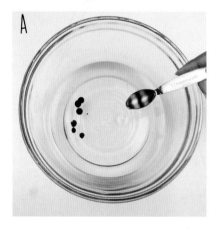

A

1 Place the glue in a large bowl. You can add coloring and/or fragrance oil at this point; note that adding fragrance oil may make your slime cloudy (**A**). Mix thoroughly. To make the sample slime, I used red and blue food coloring. Set aside. For the clearest possible slime, mix slowly. If you mix quickly, you'll need to wait longer for the slime to clear because there will be more air bubbles.

BASIC CLEAR GLUE SLIME RECIPES

- Color-Changing Slime (page 56)
- Crunchy Charms Slime (page 66)
- Fizz Slime (page 72)
- Floam (page 64)
- Hybrid Slime: Charm + Fishbowl (page 78)
- Hybrid Slime: Cloud + Clear (page 81)
- Hybrid Slime: Fizz + Inflating (page 80)
- Hybrid Slime: Fizz + Slushie (page 79)
- Hybrid Slime: Inflating + Clear Glue Floam (page 83)
- Hybrid Slime: Inflating + Icee (page 86)
- Hybrid Slime: Jelly + Clear Glue Floam (page 82)
- Hybrid Slime: Jelly Cube + Glossy (page 84)
- Jelly Cube Slime (page 68)
- Metallic Foil Slime (page 70)
- Pompom Slime (page 58)

BASIC CLEAR GLUE SLIME PROJECTS

- Brushing on Pigment (page 90)
- Extreme Avalanche Slime (page 96)
- Floam Explosions (page 102)
- Slime and Squishies (page 94)
- Slime Designs (page 92)
- Slime Palettes (page 100)
- Slime Sculptures (page 104)

B

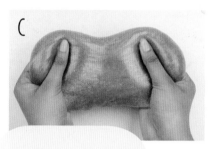

C

Yield: Approximately 14 ounces (410 g)

2 Add your activator to the glue in small amounts, no more than 2 tablespoons (28 ml) at a time (B). See the guidelines below for the recommended amounts of each type of activator—be sure to add only one type of activator to each batch of slime. Stir the mixture well after adding the activator to make sure you don't add too much. Also, the amount you add will affect the slime's consistency, and you can adjust it to achieve a consistency you like.

3 Once the mixture is slightly sticky, start to knead the slime (C). Dip your fingers in the activator before kneading so less slime will stick to your hands. Playing with slime is the best way to mix it fully and achieve

the best possible texture. Remember to have clean hands when doing this, as anything on your hands will be seen in the clear slime.

4 Store the finished slime in an airtight container so it doesn't dry out. When the slime is freshly made, it will be cloudy due to air bubbles incorporated in during the mixing process. After letting the slime sit for 3 to 5 days without touching it, it will clear up. The slime will be fully clear if you use contact lens solution and baking soda, but will be cloudy if you use liquid starch or laundry detergent as an activator. After you let it sit, the slime should also rip less as there aren't as many air bubbles.

ACTIVATOR	RECOMMENDED AMOUNT
Liquid laundry detergent	4 to 6 tablespoons (60 to 90 ml)
Liquid starch (such as Sta-Flo)	5 to 7 tablespoons (75 to 105 ml)
Baking soda + contact lens solution	Add 1 tablespoon (15 ml) of baking soda to the glue mixture and then add 1 to 3 tablespoons (15 to 45 ml) of contact lens solution. Renu Fresh Multipurpose Solution works well; if you use other brands, you may need to add a lot more. I also recommend adding 1 teaspoon (5 ml) of glycerin to the glue mixture if you use this activator, although this is optional. Slime made with this activator tends to develop a hard layer on top after a week, so to combat this, I recommend adding glycerin, which keeps it stretchy longer.

BASIC CLEAR GLUE SLIME: HOW MUCH ACTIVATOR DO I NEED?

It depends on which activator you choose! See the recommended amounts below.

Add no more than 2 tablespoons (28 ml) of activator at a time and then stir well. Check the consistency of the slime before adding more. The amount of activator you add will affect the consistency; you can adjust it to a consistency you like, or one you want for a specific recipe.

If you add too much activator, your slime will become hard, but if you don't add enough, it will be too sticky.

If you make a mistake and need to fix your slime, go to the troubleshooting section on pages 18–21.

Remember—use only one type of activator for each batch of slime!

MAKE IT RIGHT: FIXES FOR STICKY SITUATIONS

SLIME WITH A TOUGH TOP LAYER

Are you having issues with slime that's been activated with contact lens solution and baking soda? Is your slime developing a tough top layer?

After much experimentation, I've found that slimes activated with contact lens solution and baking soda don't last as long as slimes made with other activators, like liquid starch. These slimes develop a tough layer on top, causing the slime to not last as long. I've experimented with many different baking-soda-to-contact-lens-solution ratios and have found the sweet spot that reduces the tough layer, which is what's listed in the two base recipes.

When making slime activated with contact lens solution and baking soda, I've found that adding in a small amount of glycerin to the glue mixture has helped combat the tough layer. I recommend adding 1 to 2 teaspoons (5 to 10 ml) only with freshly made slime. If you come back to your slimes after a couple

days and find a tough layer, add in a teaspoon (5 ml) of glycerin and knead your slime to get rid of the top layer. This technique will work several times before the slime "dies" and is no longer playable. I've also found that the slime doesn't develop as much of a top layer when I play with the slime consistently, every day. I hope this helps you to make better slimes with contact lens solution and baking soda.

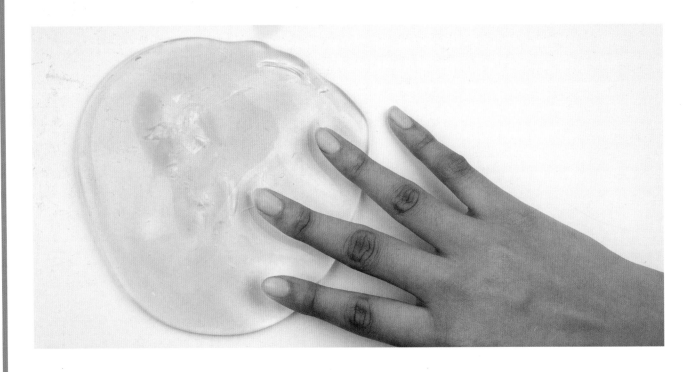

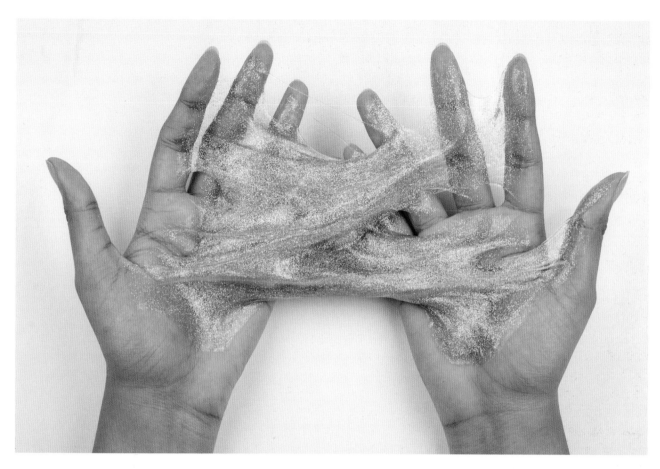

STICKY SLIME?

Has your slime liqueified after letting it sit? Do you have a sticky mess you just can't play with?

All you need to do is add more activator in to fix it! This generally fixes most slimes. If your activator isn't working, try adding in a stiff slime to thicken up the mixture. Please keep in mind that some liquefied slimes simply won't activate again no matter how much activator you add, and you'll need to throw it out.

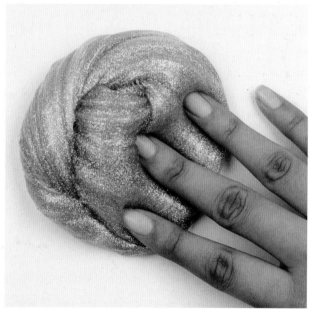

HARD OR STIFF SLIME?

Is your slime overactivated from adding in too much activator? Is it stiff from playing with it all day?

Before fixing your slime, I recommend letting it sit for 1 to 2 days. Through the process of making slime, you incorporate a lot of air bubbles. This is due to the mixing and kneading. The added air will cause slime to tear. But, over the course of 1 to 2 days, the bubbles will rise to the top of the slime and pop, leaving no air bubbles. The slime should be much stretchier after this. If the slime still isn't stretchy enough for your taste, I recommend adding lotion to white glue–based slimes. Don't add lotion to clear glue–based slimes, as this will make the slime translucent, instead of clear.

Another option would be to dip your slime in warm water, but be careful that the water isn't too hot. Please have an adult helping with this. Be careful not to add in too much water to your slime, as this will cause the slime to jiggly and even liquefy. Warm water works for both clear glue– and white glue–based slimes, as it won't affect the clarity of clear slimes.

CLEANING UP AND STORING SLIME

CLEANING UP

If small amounts of slime get stuck to clothes or carpet, or any sort of fabric or fiber, I would recommend using a basic baby wipe (any brand is fine) to rub off the slime. This has always worked well for me.

For bigger slime messes, try to get as much as you can out of the fabric, and then treat the fabric with a stain remover and wash it in the washing machine according to the garment or item's label guidelines.

If slime gets into carpet, immediately dampen the carpet and try to get as much of the slime out without rubbing the slime into the carpet. Use a carpet stain remover cleaner after that to get rid of any remaining stains.

I use a very diluted bleach solution to clean my desk to remove any stains from food coloring and any debris left from the slime-making process.

For stubborn dried-on slime that's stuck to a container, I soak the container in warm water and the slime can be easily removed.

STORING

It's important to store slime in an airtight container because it will dry out if you don't store it properly! Examples of containers I use are old yogurt/other food containers that are washed and dried—don't forget to label that they are slime. You could also use any sort of resealable bags, but I've found that this makes slime liquefy faster.

Also, always make sure to keep slime out of the reach of young children.

I also love decorating my slime containers with a variety of materials, such as washi tape, yarn, stickers, and labels. You can get creative with this and have your container match the slime within, so when you display it the slime and container both look gorgeous!

2

Add-ins for Epic Slime

THIS CHAPTER IS ALL ABOUT AWESOME ADD-INS TO SLIME THAT CAN CREATE UNIQUE AND FUN TEXTURES. THIS INCLUDES EVERYTHING FROM FRAGRANCE OILS TO SUPPLIES FOR MAKING YOUR OWN MIX-INS, WHICH YOU CAN FIND ONLINE OR IN YOUR LOCAL CRAFT STORE.

MUST-HAVE MIX-INS

THESE ARE THE ADDITIVES I USE REGULARLY TO ADD COLOR AND DIMENSION TO MY SLIME.

COLORANTS

There are a variety of ways you can add color to your slime. I like using liquid food coloring because it's easy to mix to create a variety of colors and has no detectable scent. The color shows up well in clear glue–based slime, but will appear more pastel-like in white glue–based slimes, unless you add many, many drops. I try to avoid adding too many drops as the color can stain your hands. Gel food coloring can work as well, but I find that tends to be messier.

Acrylic paint is another option to color your slime, but don't use too much as it has a scent that many people dislike, and it might dry out your slime. Note that acrylic paint shows up more vibrantly than food coloring in white glue–based slime, so if you want, for example, a vibrant red in white glue–based slime, I would recommend using the acrylic paint.

PIGMENTS

You can also add pigments to slime. They come in a wide variety, like glow in the dark, metallic, iridescent, color shifting, temperature shifting, and ones that react to UV light. For the best results, use pigments with clear glue–based slimes, where they'll show up most clearly. I generally don't recommend adding them to white glue–based slime because they aren't vibrant. In most cases you don't need too much pigment when adding it to a clear slime—a little goes a long way—but with the temperature-sensitive pigments, you'll need to add a bit more pigment in order to see the temperature-shifting properties.

FRAGRANCE OILS

Fragrance oils add a great scent to your slimes, but be sure you're using ones that are craft grade and safe for your hands, as you're adding it to slime that will come in contact with your hands. Also make sure you don't have any allergies or sensitivities to them. I don't recommend using food-grade scents in slime because they tend to liquefy it.

The amount of scent you add really depends on the size of the batch you're making. I suggest you start by adding in a couple of drops with a dropper and then gradually add more till you reach the desired scent intensity. You can also blend together various scents to create a new scent of your own!

Be careful when adding scents to clear slime because most of them will make the slime cloudy. I recommend testing fragrance oils in small batches so you can confirm which ones won't affect the clarity of the slime. If you're planning to also add color and pigment to clear slime, it should be alright if it's a little cloudy from the scent.

GLITTER

Glitter is one of my absolute favorite add-ins. It's available in such a wide variety of colors, types, and sizes. There's regular, fine, chunky, holographic, and iridescent glitter. There are also flakes that are iridescent, which look gorgeous in slime. Make sure the glitter you're using doesn't have any sharp sides; you don't want it to poke you when you're playing with the slime.

Glitter works especially well in clear glue–based slimes because you can see the glitter better. In white glue–based slimes, the glitters are harder to see. Using glitter in a white glue–based slime could be an artistic decision, such as using fine black glitter in a white base for a slime that looks like vanilla bean ice cream.

Remember that if you add in too much glitter, it can fall out of your slime and get messy. Start with a little bit and add more. Or, if you've already added in too much glitter, add in some more slime to balance things out.

FOAM BEADS

Foam beads add such a fun pop of color and change the texture of the slime. Foam beads are small Styrofoam beads that are available in a wide variety of colors. You can purchase individual color packages, or you can get multicolor packs.

The foam beads rise up to the top of the slime when you let it sit. The more you add, the crunchier your slime will become. Don't add too many, though, because then they'll fall out when you play with the slime. A couple of foam beads can be added for a simple pop of color, or you can add a bunch to fully change the texture of the slime to a floam (foam beads + clear glue slime) that's very crunchy.

I recommend testing any beads in a small batch of slime to see if their color will bleed into the slime. You don't want to use beads whose colors bleed, as this will change the color of the base slime, and you'll end up with plain while foam beads instead of colored foam beads.

The foam beads also come in a variety of sizes. The newest type of foam bead, "marshmallow" foam beads, are larger than usual foam beads and can be crushed to small pieces of Styrofoam when playing with them.

PLASTIC BEADS: FISHBOWL, SUGAR, AND SLUSHIE

There are several types of plastic beads that can be added to slime to change the texture.

Fishbowl beads are used as vase filler beads. They're disk-shaped and can have some sharp edges. Be careful when using these in slime. They make an awesome, crunchy Fishbowl Slime!

Sugar beads are also used as vase filler beads. They're very small and cylindrical shaped. They're used to make a modified Fishbowl Slime—Sugar Slime. It creates a slime with a unique texture.

Slushie beads are used to fill stuffed animals and small bean bags. They're slightly smaller than fishbowl beads and have smooth, rounded edges. Slime made with these beads is also very crunchy and fun to play with!

NEXT-LEVEL MIX-INS

THESE ARE SOME FUN ADD-INS I USE EVERY SO OFTEN TO JAZZ UP
MY SLIME! THEY ADD A POP OF COLOR AND A UNIQUE TEXTURE!

POMPOMS

Pompoms are so adorable in slime! They are available in a wide array of colors and sizes. Be careful when adding in pompoms as, over time, they shed into the slime. Pompoms look the best in clear slime so you can clearly see them. They don't look as good in white glue–based slimes because they can just make the slime look lumpy, but feel free to experiment and see what you like!

ACRYLIC CHARMS

Acrylic charms can be purchased online. These charms are a cute addition to slimes. Make sure that the charms don't have any sharp edges, as you'll be playing with them in the slime. You can also create your slimes based on charms you find. For example, if you find a palm tree charm, you could create a blue, clear glue–based slime, representing the ocean, with brown glitters, representing sand.

METALLIC FOIL

Metallic foil is available in many different colors, the main ones being gold, silver, and copper, but there are also colors like red and blue. Metallic foil is available from craft stores in sheets or flakes. The sheets can be very satisfying to squish into slime. It shows up best in clear glue–based slimes and can appear lumpy in white glue–based slimes. Be aware that, over time, it can tint your slime green.

USING AIR-DRY CLAY AND EXPANDABLE FAKE SNOW IN NEXT-LEVEL SLIME

In my first book, *Ultimate Slime*, I used air-dry clay and expandable fake snow in several classic slime recipes.

In this book, we kick things up a notch. Turn to Chapter 3, "Secrets of Clay and Snow Slime Revealed!," to learn about exciting new recipes and my most helpful tips for using air-dry clay and expandable fake snow in slime.

MAKE YOUR OWN MIX-INS

HERE ARE A BUNCH OF UNIQUE AND CUSTOMIZABLE SLIME-MAKING MIX-INS YOU CAN MAKE YOURSELF FROM UNEXPECTED HOUSEHOLD AND CRAFTING ITEMS!

FAKE CEREAL

These little foam pieces in donut- and cereal-looking shapes are made from the most surprising materials—foam hair curlers and foam darts and bullets—which can be found online or at your local dollar store.

WHAT YOU'LL NEED

Ingredients

Foam hair curlers or foam darts or bullets
Scissors

FOAM DARTS AND CURLERS

Foam darts generally come in one uniform size that's perfect for Cereal Slime. Foam hair curlers come in a variety of widths and all these widths work for slime and are very fun to play with! Try to find foam hair curlers with a more porous/textured surface instead of the smooth ones, as I find the smooth ones fall out of the slime more.

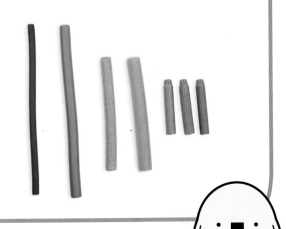

1 If using a foam dart, all you need to do is cut the dart into slices with your scissors—I prefer about ¼ inch (6 mm) thickness. You don't want to try and cut them too thin and create uneven pieces, but you also don't want to cut them too thick because then they'll be too bulky in the slime.

2 If you're using a foam hair curler, first take out the wire from the inside and then follow the same instructions as the foam dart. (A)

A

FAKE SPRINKLES AND CHOCOLATE CHUNKS

Fake sprinkles are such a cute and colorful addition to slime! You can create any color combination you choose. You can also make your own chocolate chunks out of clay that are very realistic. It's very important to make sure that neither you nor anyone else ingests them. These decorations can also be purchased on places like Ebay and Etsy, but making them yourself allows you to create your own custom color blends!

WHAT YOU'LL NEED

Ingredients

Polymer clay (I recommend Original Sculpey, as it's the softest clay)
Sharp cutting tool (blade)
Aluminum foil
Baking sheet (used for crafts only—never food)
Oven

Optional

Clay extruder
Baby wipes

1. Condition your clay. This means that you need to knead the clay. The heat from your hands will soften the clay after about 5 minutes and then you should be able to use it. The time it takes to condition your clay will vary depending on the brand. For example, if you use Original Sculpey, a beginner clay, the clay will be fairly easy to knead and become pliable, but brands like FIMO can be difficult to knead.

2. Roll out the clay into long snakes with the palm of your hand (A). You can create them any thickness that you desire, but make sure it's not too thick, as this will be difficult to cut up. Instead of rolling out the clay by hand, you could use a clay extruder (B). If you're making chocolate chunks, create the snake thicker than the one you would create for sprinkles.

3 Repeat steps 1 and 2 for all your desired colors. I recommend wiping off your hands with a baby wipe in between colors so that the color doesn't transfer.

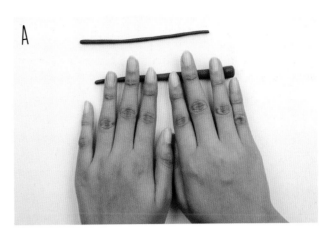

4 Place all the clay snakes on a baking sheet that's covered in aluminum foil. This baking sheet should be used for crafts only after this because it's no longer food safe once you bake polymer clay on it. You can find inexpensive baking sheets at your local dollar store.

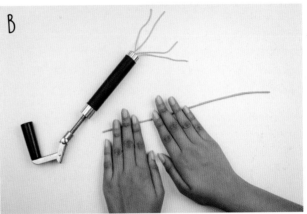

5 Bake the clay according to the instructions on the packaging. Keep an eye on the clay snakes because they're very thin and can burn easily.

6 Take the clay snakes out of the oven and let them cool.

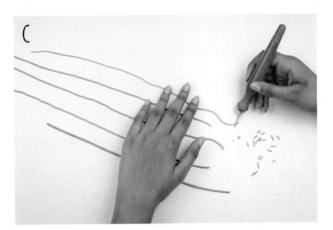

7 Use a blade to chop up the snake into whatever length sprinkles you choose (C). If you're making chocolate chunks, chop up the snake of clay at irregular angles (D). Cut them on a cutting mat meant for crafting, a piece of cardboard, or a sturdy surface you don't mind cutting into.

CHARMS

Creating your own charms for slime is so much fun because you can create any charm that you can think of! Using polymer clay, an oven-bake clay, you can create so many different charms. I am just giving one example of Twisted Marshmallow Charms because they are simple to make and you can make many of them at once, but feel free to make charms as simple or as complicated as you'd like!

WHAT YOU'LL NEED

Ingredients

Polymer clay
Sharp cutting tool (blade)
Aluminum foil
Baking sheet
Oven

Optional

Baby wipes

1 Condition your clay. This means that you need to knead the clay. The heat from your hands will soften the clay after about 5 minutes and then you should be able to use it. Depending on the brand of clay you use, the time it takes to condition your clay will vary. For example, if you use Original Sculpey, the clay will be fairly easy to knead and become pliable, but harder brands like FIMO can be difficult to knead.

2 Roll out the clay into long snakes with the palm of your hand (A). You can create them any thickness that you want.

3 Repeat steps 1 and 2 for all your desired colors. Don't forget to wipe off your hands with a baby wipe in between colors so that the color doesn't transfer.

4 Roll the snakes together and chop them up into sections using your blade (B).

5 Bake the clay according to the instructions on the packaging.

6 Take the charms out of the oven and let them cool.

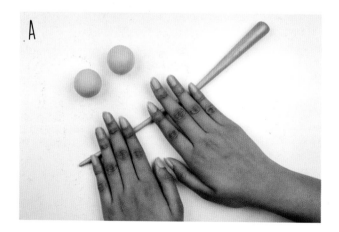

A

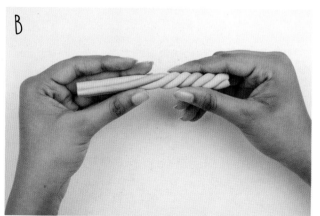

B

FOAM CUBES

These cubes are used in the Jelly Cube Slime on page 68 to create a slime with cubes that are so satisfying to squish.

WHAT YOU'LL NEED

Ingredients

Melamine sponge (chemical free)
Craft knife

1 To make these cubes, all you need to do is cut up the melamine sponges (A) into whatever size you want.

2 I usually make mine 1 inch by 1 inch by 1 inch (2.5 cm by 2.5 cm by 2.5 cm) in size and make six cubes out of one sponge (B), but you can create extra-large cubes or smaller, minicubes. The extra-large cubes might be harder to play with in slime and take longer for the cubes to fully absorb the slime, but it is an option. The minicubes are very cute, but are time-consuming to chop up.

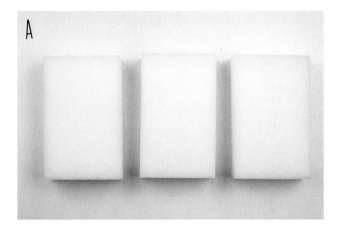

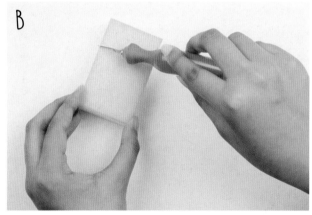

FOAM SHAPES

These foam shapes are unique because they rise to the top of the slime you put them in, like foam beads. You can create shapes in any color you want!

WHAT YOU'LL NEED

Ingredients

Foam sheet
Scissors

Optional

Ruler
Pencil
Craft Punch

1. If you want to, use a ruler to draw lines on the sheets with pencil so you can cut straight squares **(A)**.

2. Cut the foam sheet along the lines or cut the foam sheets without any guidelines **(B)**.

3. Feel free to make other shapes out of the foam sheets with scissors or use a shaped craft punch to cut them out.

A

B

FOAM CHOCOLATE CHUNKS

Using foam baby safety bumpers, you can create foam chocolate chunks that rise to the top of a slime! They're super-squishy and soft.

WHAT YOU'LL NEED

Ingredients

Foam baby safety bumpers (The product should be come flat in a large roll, not a V shape.)
Scissors

1 Cut up the baby safety bumpers into strips about 1 inch (2.5 cm) wide using scissors **(A)**. You can create smaller or larger chunks; it just depends on your personal preference.

2 Next cut along each ridge to create the individual chocolate chunk **(B)**.

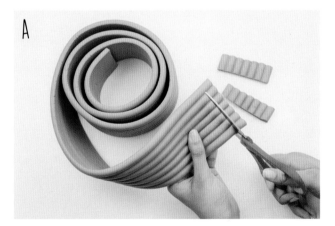

A

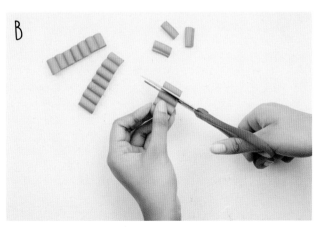

B

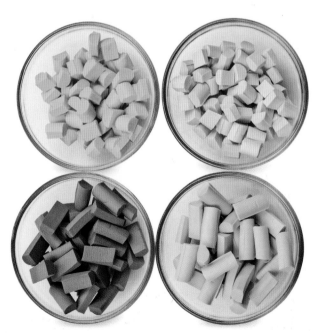

3 Secrets of Clay and Snow Slime Revealed!

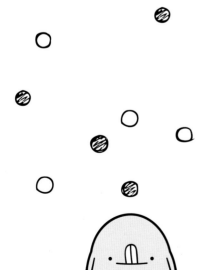

IN THIS CHAPTER, YOU'LL LEARN ALL MY BEST TIPS AND TRICKS FOR USING AIR-DRY CLAY AND EXPANDABLE FAKE SNOW IN SLIME. THERE ARE SO MANY EXCITING AND NEW RECIPES TO MAKE WITH THESE INGREDIENTS!

CLAY PLAY: ADDING AIR-DRY CLAY TO SLIME

THERE ARE MANY BRANDS OF AIR-DRY CLAY AVAILABLE, AND ALL OF THEM CREATE DIFFERENT TEXTURES WITH SLIME. THE BEST THING TO DO IS TO EXPERIMENT WITH CLAYS AVAILABLE TO YOU. I'VE DISCOVERED SOME UNIQUE CLAYS THAT I'D LIKE TO SHARE WITH YOU!

CRAYOLA MODEL MAGIC

This clay is great for creating different shapes. If you want to make a video from rolling/shaping clay, then add this to your slime. It can create a tougher slime, but this can be fixed by adding lotion. It creates a very holdable slime and is great for people with warm hands or younger children, although it can rip rather easily (A).

SMALL-QUANTITY COLOR ASSORTMENTS

These small amounts of clay, which are sold in plastic packets or containers at some retailers and online by various sellers, come in a variety of amounts (usually around ½ oz/14 g) and colors. In my experience, they're very soft and smooth, so the slime they make isn't dense, but rather a smooth, Inflating Slime. They can also create a stickier slime that's soft and soothing (B).

DAISO CLAY

This clay is very soft and creates a very soft, fluffy slime. It doesn't mold well into shapes, but is perfect for creating a nice Butter or Clay Slime. This is the clay that many slime-makers in the community use as it is easy and quick to add in (C and D).

NENDO SOFT CLAY

This clay is tougher than Daiso clay, but creates a very dense, velvety slime. This Japanese clay creates quite a unique texture (E).

When adding any clay to slime, have activator and lotion close by. Some clays, like Daiso clay, will deactivate your slime and cause it to become sticky, so you will need to activate it using activator.

Other clays, like Crayola Model Magic, become rather tough and stiff when adding it into slime. You can add lotion directly to the clay beforehand if you know the clay is stiff, or you can add it into the slime that you mixed the clay into.

Also, feel free to experiment with adding more than one type of clay to slime. For example, you can add both Nendo and Daiso clay to create a slime that's fluffy and inflating, but also velvety.

Air-dry clays for making clay slime.

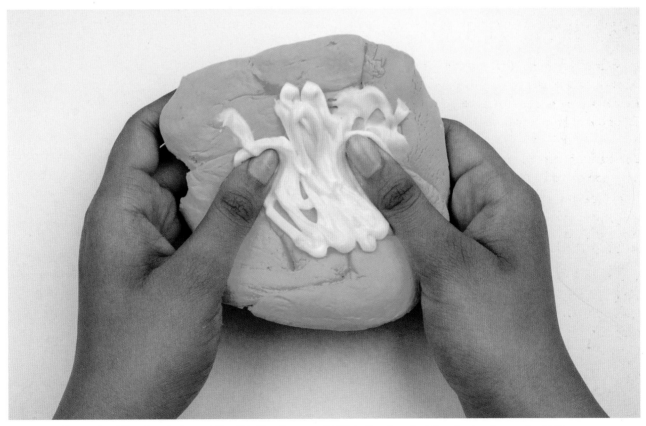

Add lotion directly to stiffer clays before adding it to the slime or add it into the slime that the clay has been mixed into.

INFLATING SLIME

AN INFLATING SLIME IS MERELY A SLIME THAT INFLATES WHEN YOU PLAY WITH IT. AIR BUBBLES GET TRAPPED IN THE SLIME, CAUSING IT TO INFLATE AND DOUBLE IN SIZE AS YOU STRETCH IT. THERE ARE MANY, MANY WAYS TO MAKE AN INFLATING SLIME. THIS IS MY PERSONAL FAVORITE BLEND OF CLAYS TO MAKE INFLATING SLIME, BUT PLEASE EXPERIMENT AND FIND A COMBINATION OF CLAYS THAT YOU LOVE!

Sunrise
Inflating Slime

WHAT YOU'LL NEED

Slime-Making Equipment

Large bowl
Mixing tool (spoon, spatula,
   or stir stick)
Measuring cups and spoons
An airtight container

Ingredients

8 ounces (235 ml) of Basic White Glue
   Slime (see page 12)
1 brick of Daiso Clay
½ brick of Nendo Clay

Optional

Color additive (see page 24)
1 to 3 drops of fragrance oil

Yield: Freshly made, the slime will be about 24 ounces (700 ml). After the slime has been in a container overnight, it will deflate, and eventually the slime will shrink to about 20 ounces (570 ml). When you play with it again, it will inflate again to the original size.

1   Make a batch of Basic White Glue Slime that's fully activated (see page 12 for recipe). Optionally, add fragrance oil when making the slime. I recommend adding the colorant after mixing in clay, as the clay's color will affect the slime's color, even if the clay is white. For example, if your base is a dark blue slime and you add white clay, this will create a light blue slime. To make the sample slime, I used yellow Daiso clay and pink Nendo clay.

2   Add the clay to the slime. For 8 ounces (235 ml) of Basic White Glue Slime, I add in 1 brick of Daiso clay and a half brick of Nendo clay (A).

B

C

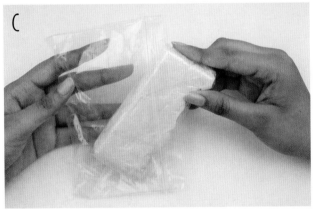

3   Mix the slime and clay together thoroughly (B).

4   If the mixture feels too sticky, add more activator. If the mixture feels too tough, add in 1 to 3 teaspoons (5 to 15 ml) of lotion.

5   Store the slime in an airtight container so it doesn't dry out (C).

VARIATIONS

Chewy Bubblegum Slime = Inflating Slime + pink coloring + bubblegum fragrance oil

Bonfire Slime = Inflating Slime + red and orange swirled coloring + summer bonfire fragrance oil

Birthday Cake Batter = Inflating Slime + light yellow coloring + sprinkles + vanilla fragrance oil

Fuzzy Blanket = Inflating Slime + plastic fake snow

Cherry Bomb = Inflating Slime + red coloring + cherry charms + cherry fragrance oil

BREAD SLIME

BREAD SLIME IS AN INFLATING SLIME THAT'S SCENTED LIKE BREAD AND IS TYPICALLY BROWN. THIS IS A SIMPLER RECIPE THAN INFLATING SLIME, USING DAISO CLAY AND A SMALL PACKET OF CLAY, BUT ANY BLEND OF CLAYS CAN BE USED TO MAKE A BREAD SLIME!

WHAT YOU'LL NEED

Slime-Making Equipment

Large bowl
Mixing tool (spoon, spatula, or stir stick)
Measuring cups and spoons
An airtight container

Ingredients

8 ounces (235 ml) Basic White Glue Slime
   (see page 12)
1 brick of white Daiso clay
1 small packet of clay
1 to 3 drops of fresh baked bread fragrance oil

Freshly Baked
Baguette Slime

A

Yield: Freshly made, the slime will be about 30 ounces (855 ml). After the slime has been in a container overnight, it will deflate, and eventually the slime will shrink to about 26 ounces (765 ml). When you play with it again, it will inflate again to the original size.

1 Make a batch of Basic White Glue Slime that's fully activated (see page 12 for recipe).

2 Add the clay to the slime (A). For 8 ounces (235 ml) of Basic White Glue Slime, I add in 1 brick of Daiso clay and 1 small packet of clay. To make the sample slime, I used white Daiso clay and a small packet of brown clay.

3 Mix the slime and clay together thoroughly (B).

B

C

 Add in 1 to 3 drops of fresh baked bread fragrance oil (C).

5 If the mixture feels too sticky, add more activator. If the mixture feels too tough, add in 1 to 3 teaspoons (5 to 15 ml) of lotion.

6 Store the slime in an airtight container so it doesn't dry out.

VARIATIONS

Strawberry Bread Slime = Light pink Inflating Slime and brown Bread Slime swirled + strawberry bread fragrance oil + strawberry charms

French Baguette Slime = Bread Slime + light brown coloring + plastic fake snow + French baguette scent

Gingerbread Slime = Bread Slime + dark brown coloring + fine black glitter + gingerbread scent

Cornbread Slime = Bread Slime + light yellow coloring + orange glitter + corn charms + cornbread scent

Blueberry Bread Slime = Light blue Inflating Slime and brown Bread Slime swirled + blueberry charms

LET IT SNOW: SLIME AND EXPANDABLE FAKE SNOW

There are two main types of expandable fake snow: SnoWonder/Instant Snow and other Super Absorbent Polymers (SAP) that have larger flakes.

These are both activated when you add water to them, meaning that they expand when water is added. Most recipes actually use dry (not hydrated) fake snow. But, for recipes that require "activated" fake snow, add equal amounts of water and SAP.

For example, if you're using 1 teaspoon (5 ml) of expandable fake snow, you would add 1 teaspoon (5 ml) of water to it. When you add more water, the fake snow turns translucent. This means that you've added too much water, and the result will be slime that leaves residue on your hands.

It's tons of fun to experiment with recipes and see what happens when you use activated expandable fake snow versus dry expandable fake snow.

The first, SnoWonder/Instant Snow, contains smaller particles. When you add water to it, it fluffs up a lot before turning translucent.

The second, SAP, contains flakier, larger pieces of fake snow, and when you add water, it doesn't fluff up that much and turns translucent with less water. I find that it's also more likely to fall out of slime when you add a lot.

SnoWonder/Instant Snow works great for Cloud Creme Slime, Icee Slime, and Cloud Slime. SAP is perfect for making thick jelly slimes, but not for other recipes.

The main issue I see with people trying to make Cloud Slime, one of the hardest slimes to make, is using SAP instead of SnoWonder/Instant Snow.

Super Absorbent Polymers (SAP)

SnoWonder/Instant Snow

Additionally, with any slime made with expandable fake snow, I recommend using liquid starch or laundry detergent as your activator. Contact lens solution and baking soda slimes tend to be a bit tougher and don't react as well with expandable fake snow. Cloud Slime is already quite difficult to make, and using an activator of contact lens solution and baking soda makes it even harder, in my experience.

Furthermore, adding fake snow may require more activator. The fake snow deactivates slime, meaning that you'll need to add in more activator than usual. On the flip side, I also find that slimes made with expandable fake snow last longer and need less reactivation.

PLASTIC FAKE SNOW AND EXPANDABLE FAKE SNOW: WHAT'S THE DIFFERENCE?

There are two types of fake snow that are used to make slime: plastic and expandable.

Plastic fake snow—the kind used to decorate for Christmas and winter holidays—consists of small flakes of plastic. When a slime recipe calls for this material, I use the brand FloraCraft, as the pieces aren't too large or sharp.

Expandable fake snow, also known as superabsorbent polymer (SAP), is made from a material that can absorb water. When you add water, it expands to a larger size. In this book, I use two types:

- **SnoWonder.** This brand has smaller pieces that work well in specific slimes.

- **SAP powder.** This comes in larger flakes, which are fun to use in some slimes but they don't work well for all slimes, such as Cloud Slime (see page 52).

For more details, check out the recipes on pages 46–53.

JELLY SLIME

THIS SLIME IS SO MUCH FUN TO PLAY WITH! IT'S PRETTY LIKE A CLEAR SLIME, BUT IT'S MUCH MORE HOLDABLE! THIS SLIME USES SAP INSTEAD OF SNOWONDER/INSTANT SNOW BECAUSE I FIND THE LARGER PARTICLES MAKE A MORE INTERESTING TEXTURE. SINCE WE AREN'T ADDING THAT MUCH FAKE SNOW, IT DOESN'T FALL OUT OF THE SLIME AS IT WOULD IF YOU WERE TO ADD MORE TO THE SLIME.

Flamingo Jelly Slime

WHAT YOU'LL NEED

Equipment

Large bowl
Measuring cups and spoons
Mixing tool (spoon, spatula, or stir stick)
Airtight container

Ingredients

1 cup (235 ml) clear PVA glue (such as Elmer's Clear Glue)
An activator (see chart on page 17 for details). I recommend using only liquid starch or laundry detergent—contact lens solution and baking soda slimes tend to be a bit tougher and doesn't react as well with expandable fake snow.
1 teaspoon (5 ml) expandable fake snow (SAP)
½ teaspoon (2.5 ml) glycerin

Optional

Color additive (see page 24)
1 to 3 drops of fragrance oil

Yield: Approximately 20 ounces (570 ml) when freshly made, but deflates to 14 ounces (410 ml)

A

1 Place the glue in a large bowl. Add 1 teaspoon (5 ml) of dry SAP and ½ teaspoon (2.5 ml) of glycerin to the glue. Optionally, add coloring additives and/or fragrance oil. To make the sample slime, I used pink food coloring. Mix thoroughly.

2 Add your activator to the glue in small amounts, no more than 2 tablespoons (28 ml) at a time (A). Stir the mixture well after adding the activator to make sure you don't add too much.

B

3 Once the mixture is slightly sticky, start to knead the slime (B). Dip your fingers in the activator before kneading so less slime will stick to your hands. Playing with slime is the best way to mix it fully and achieve the best possible texture.

4 You may need to add 1 to 2 more tablespoons (15 to 28 ml) of activator, depending on the texture of slime you like.

5 Store the finished slime in an airtight container so it doesn't dry out.

VARIATIONS

Jellyfish Jelly Slime = Pink and blue Jelly Slime swirled + blue foam beads + chunky blue glitter

Rose Water Jelly Slime = Jelly Slime + light pink coloring + light pink glitter + rose scent

Mystical Jelly Slime = Jelly Slime + dark purple coloring + chunky black glitter + gold metallic foil

Lucky Penny Jelly Slime = Jelly Slime + copper coloring + fine copper glitter + white foam beads

Midnight Jelly Slime = Jelly Slime + black coloring + silver metallic foil + fine multicolor glitter

Slurp Jelly Slime by Marjorie Lounds, @RainbowPlayMaker

ICEE SLIME

ICEE SLIME IS DIFFERENT FROM JELLY SLIME IN THAT IT HAS MORE SNOW THAN JELLY SLIME, AND IT'S ALSO SLIGHTLY MOIST AND MIGHT LEAVE A LITTLE RESIDUE ON YOUR HANDS AND TABLE. IT ALSO HAS A MUCH FLUFFIER TEXTURE THAT'S SO UNIQUE AND FUN TO PLAY WITH!

Minty Fresh
Icee Slime

WHAT YOU'LL NEED

Equipment

Large bowl
Measuring cups and spoons
Mixing tool (spoon, spatula, or stir stick)
Airtight container

Ingredients

1 cup (235 ml) clear PVA glue (such as Elmer's Clear Glue)
An activator (see chart on page 17 for details). I recommend using only liquid starch or laundry detergent—contact lens solution and baking soda slimes tend to be a bit tougher and doesn't react as well with expandable fake snow.
2½ teaspoons (12.5 ml) expandable fake snow (SnoWonder/Instant Snow)
2½ teaspoons (12.5 ml) water
½ teaspoon (2.5 ml) glycerin

Optional

Color additive (see page 24)
1 to 3 drops of fragrance oil

Yield: Approximately 22 ounces (640 ml) when freshly made, but deflates to 18 ounces (525 ml)

A

1 Add 2½ teaspoons (12.5 ml) of dry expandable fake snow (SnoWonder/Instant Snow) to a bowl. Add in 2½ teaspoons (12.5 ml) of water to the snow.

2 Place the glue in a large bowl. Add ½ teaspoon (2.5 ml) of glycerin and the expanded fake snow from step 1 to the glue. Optionally, add coloring additives and/or fragrance oil. To make the sample slime, I used green and blue food coloring. Mix thoroughly.

3 Add your activator to the glue in small amounts, no more than 2 tablespoons (28 ml) at a time (A). Stir the mixture well after adding the activator to make sure you don't add too much.

B

 Once the mixture is slightly sticky, start to knead the slime (B). Dip your fingers in the activator before kneading so less slime will stick to your hands. Playing with slime is the best way to mix it fully and achieve the best possible texture.

 You may need to add 1 to 2 more tablespoons (15 to 28 ml) of activator, depending on the texture of slime you like.

 Store the finished slime in an airtight container so it doesn't dry out.

VARIATIONS

Fresh Green Apple Icee Slime = Icee Slime + bright green coloring + green apple charms + green apple fragrance oil

Ocean Waves Icee Slime = Icee Slime + blue coloring + blue glitter + white foam beads + blue fake pearls

Mango Twist Icee Slime = Icee Slime + yellow coloring + green metallic foil + red foam beads

Eucalyptus Icee Slime = Icee Slime + light green coloring + dark green glitter + green foam beads + eucalyptus fragrance oil + small fake leaves

Strawberry Lemonade Icee Slime = Icee Slime + pink and yellow coloring + red and yellow sprinkles + a strawberry and lemon charm

CLOUD CREME

CLOUD CREME SLIME IS CLICKY AND VERY STRETCHY! CLOUD CREME SLIMES ARE SIMILAR TO GLOSSY SLIMES, BUT HAVE FAKE SNOW, WHICH GIVES THEM A SOOTHING TEXTURE. CLOUD CREME IS DIFFERENT FROM CLOUD SLIME IN THAT IT HAS LESS FAKE SNOW AND DOESN'T DRIZZLE, THOUGH IT'S POSSIBLE TO TURN A CLOUD CREME INTO A CLOUD SLIME BY ADDING MORE FAKE SNOW.

WHAT YOU'LL NEED

Equipment

Large bowl
Measuring cups and spoons
Mixing tool (spoon, spatula, or stir stick)
Airtight container

Ingredients

1 cup (235 ml) white PVA glue (such as Elmer's
   White Glue)
An activator (see chart on page 13 for details).
   I recommend using only liquid starch or laundry
   detergent—contact lens solution and baking
   soda slimes tend to be a bit tougher and doesn't
   react as well with expandable fake snow.
2½ teaspoons (12.5 ml) expandable fake snow
   (SnoWonder/Instant Snow)
1 tablespoon (15 ml) lotion
Approximately ½ cup (120 ml) foaming hand soap

Optional

Color additive (see page 24)
1 to 3 drops of fragrance oil

A

B

Yield: Approximately
18 ounces (525 ml)
when freshly made,
but deflates to 16
ounces (475 ml)

1. Place the glue in a large bowl. Add 2½ teaspoons (12.5 ml) of dry expandable fake snow (SnoWonder/Instant Snow), lotion, and foaming hand soap (A). Optionally, also add coloring additives and/or fragrance oil. To make the sample slime, I used yellow and red food coloring. Mix thoroughly.

2. Add your activator to the glue in small amounts, no more than 2 tablespoons (28 ml) at a time (B). Stir the mixture well after adding the activator to make sure you don't add too much.

Juicy Peach Cloud Crème

3. Once the mixture is slightly sticky, start to knead the slime. Dip your fingers in the activator before kneading so less slime will stick to your hands. Playing with slime is the best way to mix it fully and achieve the best possible texture.

4. You may need to add 1 to 2 more tablespoons (15 to 28 ml) of activator, depending on the texture of slime you like.

5. Store the finished slime in an airtight container so it doesn't dry out.

VARIATIONS

Vanilla Bean Cloud Creme = Cloud Creme + fine black glitter + vanilla bean fragrance oil

Mermaid Cloud Creme = Cloud Creme + purple and blue coloring swirled + mermaid tail charms + blue and purple sprinkles

Spiced Cinnamon Cloud Creme = Cloud Creme + light brown coloring + white foam chocolate chunks + dark brown fine glitter

Pineapple Party Cloud Creme = Cloud Creme + bright yellow coloring + white foam beads + pineapple charms + topped with yellow chunky glitter

Butterscotch Cloud Creme = Cloud Creme + medium brown coloring + drizzle white Glossy Slime on top + butterscotch charms + butterscotch fragrance oil

CLOUD SLIME

THIS SLIME IS THE HARDEST, MOST TIME-CONSUMING SLIME TO MAKE, BUT IT'S ALSO THE MOST REWARDING. DESPITE NEEDING TO KNEAD AND MIX THIS SLIME FOREVER, THE RESULTING DRIZZLES ARE SO MESMERIZING THAT IT'S WORTH IT. ALSO, KEEP IN MIND THAT THIS CAN BE A MESSY PROCESS.

WHAT YOU'LL NEED

Equipment

Large bowl
Measuring cups and spoons
Mixing tool (spoon, spatula, or stir stick)
Airtight container

Ingredients

1 cup (235 ml) white PVA glue (such as Elmer's School Glue)
½ cup (120 ml) water
1 tablespoon (15 ml) lotion
Approximately ½ cup (120 ml) foaming hand soap
Activator: Use only liquid starch; laundry detergent or contact lens solution and baking soda slimes don't react as well with expandable fake snow
2 tablespoons (28 ml) expandable fake snow (SnoWonder/Instant Snow)
2 tablespoons (28 ml) water
1 teaspoon (5 ml) expandable fake snow (SnoWonder/Instant Snow)

Optional

Color additive (see page 24)
1 to 3 drops of fragrance oil

Yield: Approximately 24 ounces (700 ml)

Midnight Waves Cloud Slime

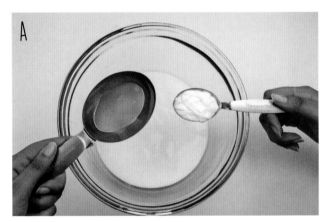

1   Place the glue in a large bowl. Add the water, lotion, and foaming hand soap (A). Optionally, also add coloring additives and/or fragrance oil. To make the sample slime, I used blue food coloring. Mix thoroughly.

2   Add the liquid starch to the glue mixture in small amounts, no more than 2 tablespoons (28 ml) at a time (B). Stir the mixture well after adding the liquid starch to make sure you don't add too much.

B

C

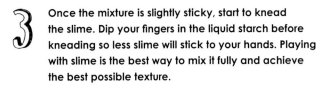
3 Once the mixture is slightly sticky, start to knead the slime. Dip your fingers in the liquid starch before kneading so less slime will stick to your hands. Playing with slime is the best way to mix it fully and achieve the best possible texture.

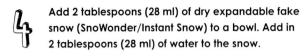
4 Add 2 tablespoons (28 ml) of dry expandable fake snow (SnoWonder/Instant Snow) to a bowl. Add in 2 tablespoons (28 ml) of water to the snow.

5 Add the expanded fake snow into the White Glue Slime from step 3 slowly, adding 1 tablespoon (15 ml) at a time and kneading (C). You may need to add more activator while mixing if the slime is sticky. Add activator in slowly, only 1 tablespoon (15 ml) at a time.

6 Store the slime in an airtight container so it doesn't dry out.

7 Wait 3 days and then check the slime to see if it drizzles. If it doesn't, add 1 teaspoon (5 ml) more of dry expandable fake snow.

VARIATIONS

Vanilla Rose Cloud Slime = Light pink and white Cloud Slime layered + red glitter + rose charms

Soothing Lavender Cloud Slime = Cloud Slime + light purple coloring + fake leaves + purple fake pearls + lavender fragrance oil

Head in the Clouds Slime = Cloud Slime + white and blue coloring swirled + silver metallic foil + cloud charm

Flamingo Feathers Cloud Slime = Cloud Slime + bright pink coloring + topped with white glitter + flamingo charms

Summer Daze Cloud Slime = Cloud Slime + yellow coloring + green foam beads + gold metallic foil

CLOUD SLIME: TROUBLESHOOTING TIPS

Why isn't my slime drizzling? Why is it so stringy?

- Let sit 3 days. Drizzle.

Why is the fake snow falling out? Why is it leaving a residue on my table/hands?

- If you see a little fallout when you drizzle, you may need to let it sit longer and add a small amount of activator.

- If you get fake snow on your fingers when you stick them into the container, add a dollop of clear slime.

Is there any way to make this slime faster?

- Pull the slime—it incorporates the snow better.

- Add 2 more teaspoons (10 ml) of fake snow with a little water in it.

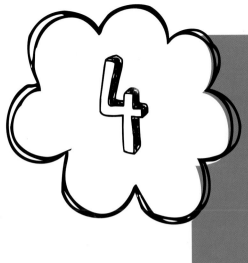

4 Extreme Slime Recipes

IN THIS CHAPTER, YOU'LL
LEARN TO MAKE NEW SLIMES
USING FUN ADD-INS!

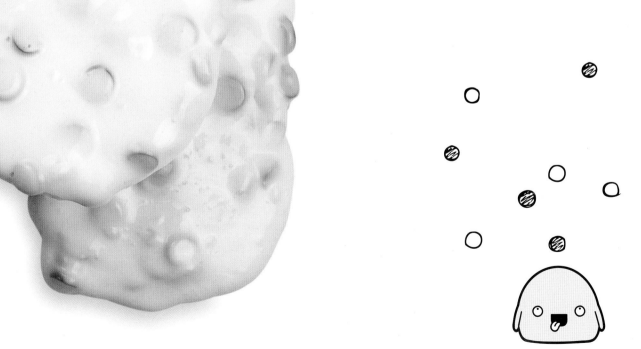

COLOR-CHANGING SLIME

THIS SLIME IS INTERESTING AND UNLIKE ANY OTHER SLIME IN THE BOOK! PLAYING WITH THIS SLIME IS LIKE MAGIC. THE MOST IMPORTANT PART OF THIS SLIME IS THE PIGMENT, SO MAKE SURE YOU CHECK PAGE 24 FOR MORE DETAILS ON THE COLOR-CHANGING PIGMENT NEEDED FOR THIS SLIME! THIS RECIPE USES COLOR-CHANGING PIGMENT THAT REACTS TO HEAT AND COLD, BUT CHECK OUT OTHER PIGMENTS, SUCH AS ONES THAT GLOW IN THE DARK OR CHANGE COLOR WITH UV LIGHT TO MAKE OTHER COLORFUL SLIMES.

Magenta Magic Color-Changing Slime

WHAT YOU'LL NEED

Slime-Making Equipment

Large bowl
Mixing tool (spoon, spatula, or stir stick)
Measuring cups and spoons
An airtight container

Ingredients

8 ounces (235 ml) Basic Clear Glue Slime (see page 16)
1 tablespoon (15 ml) of color-changing pigment

Optional

1 to 3 drops of fragrance oil

Yield: 8 ounces (235 ml)

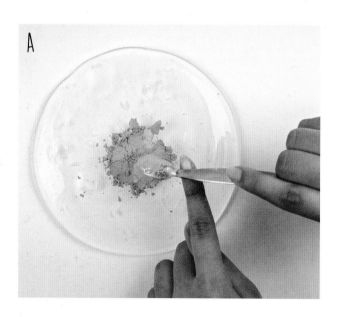

A

1 Make a batch of Basic Clear Glue Slime that's fully activated and not sticky (see page 16 for recipe). Optionally, add fragrance oil when making the slime. Be sure to use a fragrance oil that's light in color so it doesn't discolor your slime.

2 Add 1 tablespoon (15 ml) of color changing pigment to the slime (A). Depending on the brand of pigment you purchase, you may need to add more or less to achieve the desired color. I recommend adding 1 tablespoon (15 ml) to start and then adding more till it is an opaque solid color.

3 Mix the slime and pigment together thoroughly (B).

B

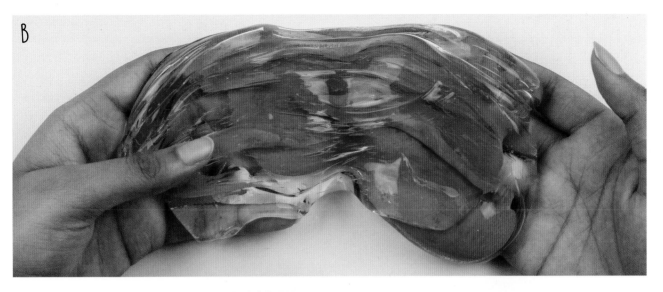

C

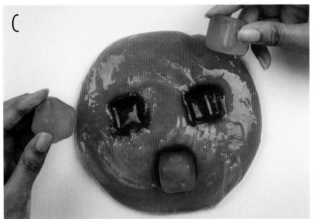

D

 Play with the slime! Depending on the pigment you purchase, the slime may change color with cold or heat. For pigments that react to the cold, I recommend using a cold water bottle from your fridge or an ice cube for even better effects (C). For heat, I recommend using a blow dryer or warmed spoons to change the color of the slime if the heat from your hands, alone, doesn't change the color of the slime (D).

 Store the slime in an airtight container so it doesn't dry out.

VARIATIONS

Planet Earth Color-Changing Slime = Basic Clear Glue Slime + green-to-blue color-changing pigment + small people charms

Groovy Color-Changing Slime = Basic Clear Glue Slime + purple-to-turquoise color-changing pigment + black glitter

Pink Candy Color-Changing Slime = Basic Clear Glue Slime + black-to-pink color-changing pigment + slushie beads + candy scent

Caribbean Salsa Color-Changing Slime = Basic Clear Glue Slime + red-to-yellow color-changing pigment + green foam beads + tropical scent

Oreo Color-Changing Slime = Basic Clear Glue Slime + vanilla-to-black color-changing pigment + Oreo scent

POMPOM SLIME

THIS SLIME IS SO ADORABLE! POMPOMS MAKE THE BEST ADDITION TO SLIME. THERE ARE ENDLESS POSSIBILITIES WITH POMPOMS BECAUSE THEY'RE AVAILABLE IN SO MANY DIFFERENT COLORS AND VARIATIONS. KEEP IN MIND THAT EXTENDED PLAY WITH THIS SLIME WILL CAUSE THE POMPOM'S FIBERS TO SHED, SO IT DOESN'T MAINTAIN ITS CLARITY AS LONG AS OTHER CLEAR GLUE-BASED SLIMES.

Pompom
Dreams
Slime

WHAT YOU'LL NEED

Slime-Making Equipment

Large bowl
Mixing tool (spoon, spatula, or stir stick)
Measuring cups and spoons
An airtight container

Ingredients

One batch of Basic Clear Glue Slime (see page 16)
Pompoms

Optional

Color additive (see page 24)
1 to 3 drops of fragrance oil

Yield: Depends on the amount of slime and pompoms you use

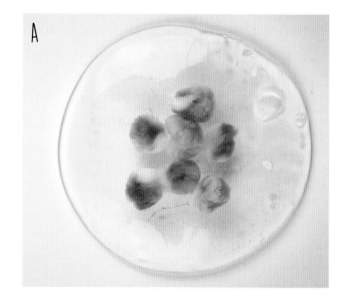

A

Make a batch of Basic Clear Glue Slime that's fully activated (see page 16 for recipe). The more pompoms you add, the less stretchy your slime will be and the tougher the slime will become as the pompoms will absorb some of the slime. Thus, the more pompoms you add, the stickier you want the slime to be, as the slime needs to be able to hold the pompoms. Optionally, add fragrance oil and/or coloring when making the slime.

POMPOM SIZE

The size of your pompoms will affect how many you add. I prefer to add about a third of pompoms as slime (by volume), regardless of the pompoms' size.

B

2 Add in pompoms to the slime (A). I would recommend starting with a small amount (3 to 4 pompoms) and playing with the slime and then adding more (1 to 2 at a time) till you reach the desired texture.

3 Mix the slime and pompoms together thoroughly (B).

4 If the mixture feels too sticky, add more activator.

5 Store the slime in an airtight container so it doesn't dry out.

Shinjuku Pop Slime by Erin Lutterbach Murphy, The Slimeonade Stand

VARIATIONS

Aloha Slime = Basic Clear Glue Slime + blue coloring + pink and green pompoms + yellow glitter

Candy Crush Slime = Basic Clear Glue Slime + multicolor pompoms

Autumn Slime = Basic Clear Glue Slime + amber coloring + green, yellow, and red pompoms + leaf charms

Jelly Bean Slime = Basic Clear Glue Slime + multicolor pompoms + multicolor glitter

Green Jolly Rancher Slime = Basic Clear Glue Slime + green pompoms + green apple scent

CHOCOLATE CHUNK SLIME

THIS SLIME—ALSO KNOWN IN THE SLIME COMMUNITY AS JAVA CHIP SLIME—LOOKS JUST LIKE IT HAS LITTLE CHOCOLATE CHUNKS FLOATING ON TOP! WHEN YOU LET THIS SLIME SIT OVERNIGHT, THE FOAM CHUNKS RISE TO THE TOP—IT'S VERY SATISFYING TO MIX THEM BACK IN!

WHAT YOU'LL NEED

Slime-Making Equipment

Large bowl
Mixing tool (spoon, spatula, or stir stick)
Measuring cups and spoons
An airtight container

Ingredients

One batch of Basic White Glue Slime (see page 12)
Chocolate Chunks (see page 30)—I would recommend using enough chunks to create a thin layer on the top of the slime, so for 8 ounces (235 ml) of slime, use about 8 to 10 chunks. This number may vary depending on what size you choose to cut you chocolate chunks.

Optional

Color additive (see page 24)
1 to 3 drops of fragrance oil

Yield: Depends on the amount of slime and Chocolate Chunks you use

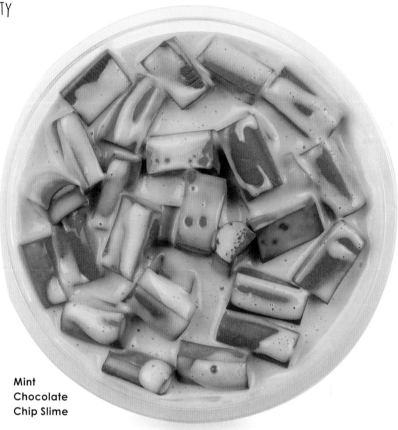

Mint Chocolate Chip Slime

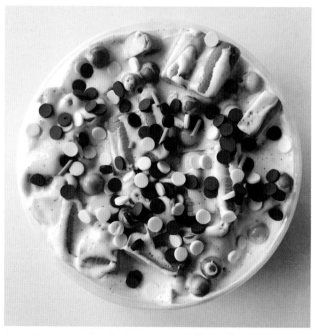

Java Chunk Slime by Chelsey P., H0nestslimereviews and H0nestlyslime

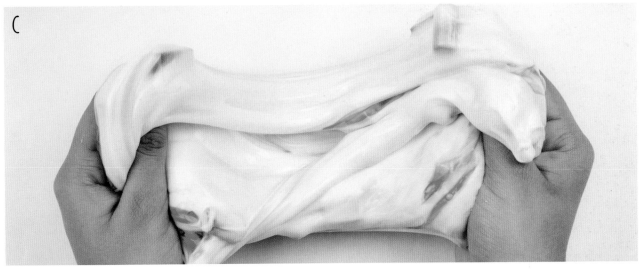

1 Make a batch of Basic White Glue Slime that's fully activated (see page 12 for recipe).

2 Add Chocolate Chunks to the slime; optionally, add coloring and/or fragrance (A). To make the sample slime, I used blue and green food coloring. I would recommend starting with a small amount (3 to 4 chunks) and playing with the slime and then adding more (1 to 2 at a time) till you reach the desired texture. If you add too many chunks, the slime becomes difficult to play with as it rips (B).

3 Mix the slime and Chocolate Chunks together thoroughly (C).

4 Store the slime in an airtight container so it doesn't dry out.

VARIATIONS

Brownie Batter Slime = Basic White Glue Slime + light brown coloring + dark brown chocolate chunks + brownie fragrance oil

S'mores Slime = Basic White Glue Slime + beige coloring + dark brown and white chocolate chunks

Chocolate Chip Cookie Dough Slime = Basic White Glue Slime + beige coloring + brown chocolate chunks + cookie fragrance oil

Peppermint Slime = Basic White Glue Slime + red chocolate chunks + fine red glitter

Red Velvet Slime = Basic White Glue Slime + red coloring + white chocolate chunks + heart charms

CEREAL SLIME

THIS SLIME LOOKS JUST LIKE A BOWL OF CEREAL! THESE MINI DONUT-SHAPED PIECES ARE ADORABLE AND MAKE A GREAT ADDITION TO SLIME. YOU CAN MAKE THIS SLIME REALISTIC OR, AS I PREFER, USE ANY COLOR YOU'D LIKE! WHEN YOU LET THIS SLIME SIT OVERNIGHT, THE CEREAL FOAM PIECES RISE TO THE TOP, AND IT'S SO FUN TO CRUNCH AND SQUISH THEM BACK INTO THE SLIME!

Mermaid Cereal Slime

WHAT YOU'LL NEED

Slime-Making Equipment

Large bowl
Mixing tool (spoon, spatula, or stir stick)
Measuring cups and spoons
An airtight container

Ingredients

One batch of Basic White Glue Slime
   (see page 12)
Cereal Foam Pieces (see page 29)
I would recommend using enough Cereal Foam
   Pieces to create a layer on the top of the slime,
   so for an 8-ounce (235 ml) slime, about 12 to
   15 pieces, but this might vary depending on
   what size you choose to cut your pieces.

Optional

Color additive (see page 24)
1 to 3 drops of fragrance oil

Yield: Depends on
the amount of slime
and cereal foam
pieces you use

A

B

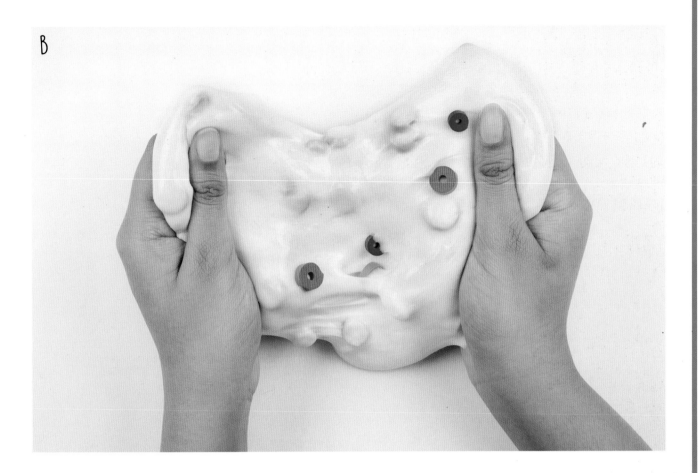

1 Make a batch of Basic White Glue Slime that's fully activated (see page 12 for recipe).

2 Add in cereal foam pieces to the slime (A). I recommend beginning with a small portion and mixing it into the slime and then adding some more (1 to 2 at a time) till you reach the desired texture. If you add too many cereal foam pieces into your slime, the slime becomes tougher to play with because it rips.

3 Mix the slime and cereal foam pieces together thoroughly (B).

4 Store the slime in an airtight container so it doesn't dry out.

VARIATIONS

Fruit Loops Slime = Basic White Glue Slime + multi-color cereal foam pieces + multicolor fine glitter

Cheerios Slime = Basic White Glue Slime + yellow coloring + beige cereal foam pieces + honey fragrance oil + bee charms

Cinnamon Roll Slime = Basic White Glue Slime + light brown cereal foam pieces + dark brown glitter + cinnamon fragrance oil

Pop of Pastel Slime = Basic White Glue Slime + pastel yellow coloring + pastel pink, blue, and green cereal foam pieces

Candy Corn Slime = Basic White Glue Slime + white, orange, and yellow cereal foam pieces + candy corn charms

FOAM SHAPES SLIME

THIS SLIME IS SO FUN TO CUSTOMIZE AND MAKE YOUR OWN! YOU CAN CUT YOUR FOAM PIECES INTO ANY SHAPE YOU LIKE, AND THEY CAN BE ANY COLOR IMAGINABLE. SINCE THE ADD-IN IS MADE OF FOAM, WHEN YOU LET THIS SLIME SIT OVERNIGHT, THE FOAM PIECES WILL RISE UP TO THE TOP AND CREATE A LAYER THAT YOU CAN SQUISH BACK IN! NOTE THAT THESE FOAM PIECES DON'T BREAK DOWN INTO THE SLIME.

WHAT YOU'LL NEED

Slime-Making Equipment

Large bowl
Mixing tool (spoon, spatula, or stir stick)
Measuring cups and spoons
An airtight container

Ingredients

One batch of Basic White Glue Slime (see page 12)
Square Foam Pieces, or any other shape you make (see page 34)—I recommend using enough square foam pieces to create a layer on the top of the slime, so for an 8 ounces (235 ml) slime, about 12 to 14 pieces, but this might vary depending on what size and shape you choose to cut your pieces.

Optional

Color additive (see page 24)
1 to 3 drops of fragrance oil

Yield: Depends on the amount of slime and foam shapes you use

Strawberry Checkerboard Slime

A

| Make a batch of Basic White Glue Slime that's fully activated (see page 12 for recipe).

B

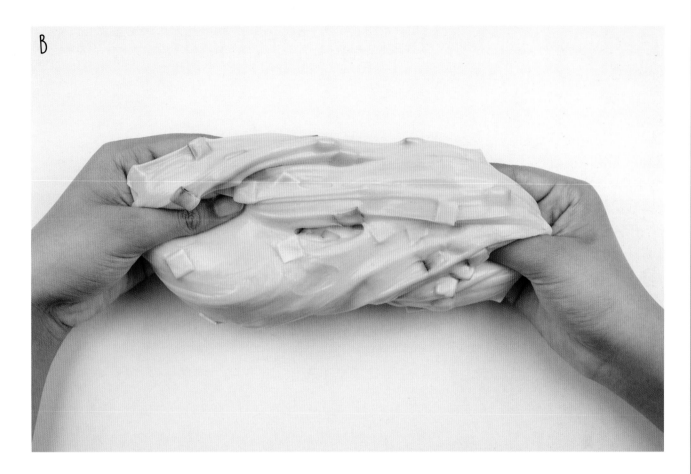

 Add foam shapes to the slime; optionally, add coloring and/or fragrance (A). To make the sample slime, I used pink food coloring. I recommend beginning with a small portion of shapes and mixing it into the slime and then adding some more (1 to 2 at a time) till you reach the desired texture. If you add too many foam shapes into your slime, the slime becomes tougher to play with because it rips.

Mix the slime and foam shapes together thoroughly (B).

Store the slime in an airtight container so it doesn't dry out.

VARIATIONS

Starry Night Slime = Basic White Glue Slime + dark blue coloring + white, yellow, and light blue square foam pieces

Cherry Jolly Rancher Slime = Basic White Glue Slime + red circular foam pieces + cherry scent

Snowed-in Slime = Basic White Glue Slime + blue square foam pieces + snowflake glitter + white foam beads

Candy Land Slime = Basic White Glue Slime + multicolor circular foam pieces + multicolor foam beads

Valentine's Slime = Basic White Glue Slime + pink coloring + red and white foam pieces + dark pink sprinkles + chocolate fragrance oil

CRUNCHY CHARMS SLIME

THERE ARE ENDLESS POSSIBILITIES WITH THIS CRUNCHY SLIME, AS THERE ARE SO MANY DIFFERENT CHARMS AND COMBINATIONS YOU CAN CREATE! WHEN YOU LET THE SLIME SIT, THE CHARMS WILL SINK TO THE BOTTOM AND THE SLIME WILL RISE UP TO THE TOP. THIS CAN SOMETIMES MAKE IT DIFFICULT TO GET THE SLIME OUT OF THE CONTAINER, SO I RECOMMEND USING A METAL SPOON TO HELP TAKE IT OUT!

Pop of Pearls Slime

WHAT YOU'LL NEED

Slime-Making Equipment

Large bowl
Mixing tool (spoon, spatula, or stir stick)
Measuring cups and spoons
An airtight container

Ingredients

One batch of Basic Clear Glue Slime (see page 16)
Acrylic Charms (see page 27)—the size of your acrylic charms will change how many you add. As a general rule, I would say use two-thirds of the amount of charms as slime (by volume), but that is my personal preference as it creates a crunchy slime that's also stretchy and fun to play with.

Optional

Color additive (see page 24)
1 to 3 drops of fragrance oil

Yield: Depends on the amount of slime and charms you use

C

1. Make a batch of Basic Clear Glue Slime that is slightly sticky and stretchy (see page 16 for recipe). The more charms you add, the less stretchy your slime will be, but it will become crunchier. Thus, the more charms you add, the stickier you want the slime to be because the slime needs to be able to hold the charms. On the flip side, if you add fewer charms, the slime will be more stretchy but less crunchy. Optionally, add fragrance oil and/or coloring when making the slime.

2. Add in charms to the slime (A). I would recommend starting with a small amount (3 to 4 charms) and playing with the slime and then adding more (1 to 2 at a time) till you reach the desired texture.

3. Mix the slime and charms together thoroughly (B & C).

4. If the mixture feels too sticky, add more activator.

5. Store the slime in an airtight container so it doesn't dry out.

Clear Gloss with Charms by Marjorie Lounds, @RainbowPlayMaker

VARIATIONS

Willy Wonka Slime = Basic Clear Glue Slime + gold coloring + chocolate bar charms + red foam beads

Ocean Cruise Slime = Basic Clear Glue Slime + light blue coloring + whale charms + dark blue glitter

Flower Power Slime = Basic Clear Glue Slime + light green coloring + flower charms + silver foil + pastel sprinkles

Fruit Salad Slime = Basic Clear Glue Slime + fruit charms + multi-color foam beads + gold glitter

Candy Store Slime = Basic Clear Glue Slime + candy charms + silver glitter

JELLY CUBE SLIME

THIS SLIME HAS A TEXTURE THAT REALLY STANDS OUT! FIRST, THE SLIME HAS CUBES, MAKING IT VISUALLY APPEALING AND FUN TO STRETCH; THEN, ONCE YOU SQUISH THE JELLY CUBES, THEY BREAK DOWN IN THE SLIME FOR A UNIQUE, SQUISHY TEXTURE!

Peach Paradise Slime

WHAT YOU'LL NEED

Slime-Making Equipment

Large bowl
Mixing tool (spoon, spatula, or stir stick)
Measuring cups and spoons
An airtight container

Ingredients

One batch of Basic Clear Glue Slime (see page 16)
Foam cubes (see page 33)

Optional

Color additive (see page 24)
1 to 3 drops of fragrance oil

Yield: Depends on the amount of slime and cubes you use

FOAM-CUBE SIZES

The size of your cubes will affect how many you add. As a general rule, I like to add cubes that are equal to about two-thirds of the amount of slime (by volume).

A

1 Make a batch of Basic Clear Glue Slime that's fully activated (see page 16 for recipe). The more cubes you add, the less stretchy your slime will be and the tougher the slime will become, as the cubes will absorb lots of the slime. Thus, the more cubes you add, the stickier you want the slime to be, so the slime is able to hold the cubes.

2 Add the cubes to the slime; optionally, add coloring and/or fragrance (A). To make the sample slime, I used yellow and red food coloring. I recommend starting with a small amount (3 to 4 cubes) and playing with the slime and then adding more (1 to 2 at a time) till you reach the desired texture.

3 Mix the slime and cubes together thoroughly (B).

4 To achieve the texture shown below, continue to squish and squeeze the cubes (C). If the mixture feels too sticky, add more activator.

5 Store the slime in an airtight container so it doesn't dry out.

VARIATIONS

Lemon Slime = Basic Clear Glue Slime + yellow coloring + jelly cubes + lemon charms + yellow foam beads

Galaxy Goo Slime = Basic Clear Glue Slime + purple coloring + jelly cubes + blue and silver glitter

Black Tie Slime = Basic Clear Glue Slime + black coloring + jelly cubes + gold glitter

Cola Slime = Basic Clear Glue Slime + brown coloring + jelly cubes + cola fragrance oil + cola bottle charms + red foam beads

Candy Apple Slime = Basic Clear Glue Slime + red coloring + jelly cubes + red glitter

B
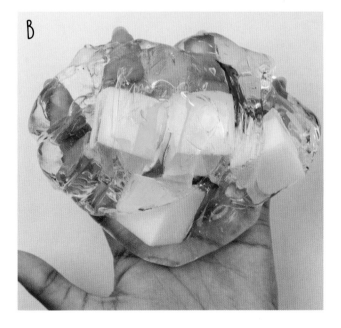

C
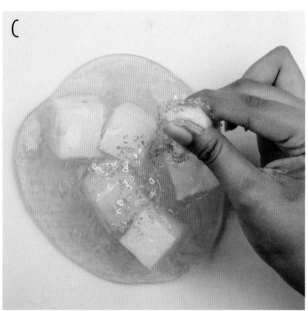

METALLIC FOIL SLIME

THIS SLIME IS SO GORGEOUS AND ELEGANT. THE FOIL IS SO PRETTY—ESPECIALLY THE FOIL SHEETS—AND IT MAKES A VERY SATISFYING SOUND AND A VISUALLY PLEASING EFFECT. NOTE THAT OVER TIME METALLIC FOIL WILL DISCOLOR YOUR SLIME AND GIVE IT A GREEN TINT, BUT IT SHOULD STAY CLEAR AND PRETTY FOR 1 TO 2 WEEKS.

Summer Lagoon Slime

WHAT YOU'LL NEED

Slime-Making Equipment

Large bowl
Mixing tool (spoon, spatula, or stir stick)
Measuring cups and spoons
An airtight container

Ingredients

One batch of Basic Clear Glue Slime (see page 16)
1 sheet of metallic foil or a pinch of flakes

Optional

Color additive (see page 24)
1 to 3 drops of fragrance oil

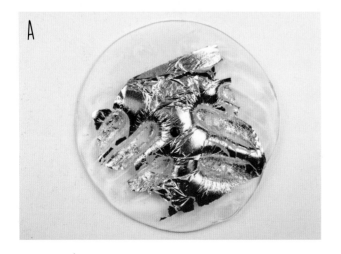

A

1 Make a batch of Basic Clear Glue Slime that's fully activated (see page 16 for recipe).

2 Measure out the amount of metallic foil (A). Keep in mind, the more foil you add, the faster the slime will discolor.

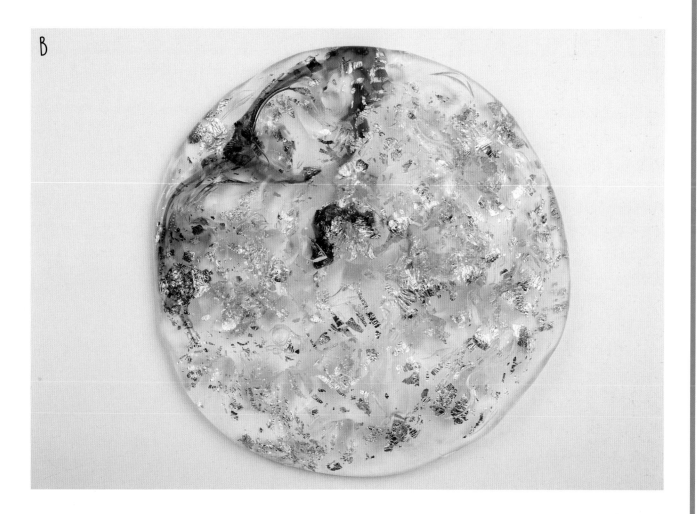

 3 Optionally, add coloring and/or fragrance. To make the sample slime, I used blue food coloring. Mix the slime and metallic foil together thoroughly **(B)**.

 4 Store the slime in an airtight container so it doesn't dry out.

Yield: 14 ounces (415 ml)

VARIATIONS

Diva Slime = Basic Clear Glue Slime + pink coloring + gold foil + chunky pink glitter

Champagne Slime = Basic Clear Glue Slime + yellow coloring + gold foil + yellow glitter + white foam beads

Frozen Slime = Basic Clear Glue Slime + blue coloring + blue and silver foil + blue sprinkles + snowflake charms

Summer Sun Slime = Basic Clear Glue Slime + orange coloring + gold foil + yellow glitter + sweet pea fragrance oil

Victorious Slime = Basic Clear Glue Slime + gold foil + chunky black glitter

FIZZ SLIME

THIS SLIME IS VERY CRUNCHY AND HAS SUCH A UNIQUE TEXTURE. THIS SLIME IS SO MUCH FUN TO CUSTOMIZE AND MAKE YOUR OWN. IN ADDITION TO BEING ABLE TO CUSTOMIZE THE COLOR, YOU CAN ALSO CHOOSE THE AMOUNT OF CRUNCHINESS YOU WANT FOR YOUR SLIME! THE FOCUS FOR THIS SLIME IS THE RATIO BETWEEN THE BASIC CLEAR GLUE SLIME BASE AND THE ADD-IN PLASTIC FAKE SNOW. THE MORE FAKE SNOW YOU ADD, THE CRUNCHIER THE SLIME BECOMES, BUT IT ALSO BECOMES LESS STRETCHY AND RIPS EASIER. EXPERIMENTING TO FIND YOUR PERSONAL PREFERENCE IS THE BEST PART OF THIS RECIPE!

Ultramarine Crunch Slime

WHAT YOU'LL NEED

Slime-Making Equipment

Large bowl
Mixing tool (spoon, spatula, or stir stick)
Measuring cups and spoons
An airtight container

Ingredients

One batch of Basic Clear Glue Slime (see page 16)
Plastic fake snow (see suggested ratios below)

Optional

Color additive (see page 24)
1 to 3 drops of fragrance oil

Yield: Depends on the amount of slime and fake snow you use

HOW MUCH PLASTIC FAKE SNOW?

When making Fizz Slime, I typically use one of following ratios of plastic fake snow to clear slime:

1. Half the amount of plastic fake snow as slime (by volume)

2. The same amount of plastic fake snow as slime (by volume)

3. Double the amount of plastic fake snow as slime (by volume)

A

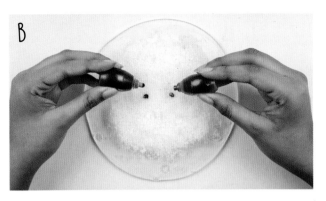

B

C

VARIATIONS

Orange Soda Fizz Slime = Clear Fizz Slime + orange coloring + soda fragrance oil

Bumblebee Fizz Slime = Layered black-and-yellow Fizz Slime + bee charms

Wicked Fizz Slime = Clear Fizz Slime + emerald green coloring + gold foil

Galaxy Pop Slime = Clear Fizz Slime + royal purple coloring + blue, silver, and purple glitter

Sea Foam Slime = Clear Fizz Slime + blue coloring + white foam beads

1. Make a batch of Basic Clear Glue Slime that's slightly sticky and stretchy (see page 16 for recipe). The more fake snow you add, the stickier you want the slime to be, so the slime can hold the fake snow.

2. Measure out the amount of fake snow you'd like to use (A). Keep in mind, the more fake snow you add, the crunchier the slime will become. Optionally, add fragrance oil and/or coloring (B). To make the sample slime, I used blue and red food coloring.

3. Mix the slime and fake snow together thoroughly (C).

4. If the mixture feels too sticky, add more activator. Feel free to add more fake snow to the slime until you reach a level of crunchiness you prefer. Remember, playing with slime is the best way to fully mix the fake snow and slime together.

5. Store the slime in an airtight container so it doesn't dry out.

Snow Fizz Slime by Chelsey P., H0nestslimereviews and H0nestlyslime

↓ RECIPE

KAWAII SLIME

"KAWAII" IS A STYLE THAT ORIGINATED IN JAPAN AND IS CENTERED AROUND ALL THINGS CUTE AND ADORABLE. A SLIME THAT IS KAWAII GENERALLY FEATURES PASTEL COLORS AND CUTE CHARMS. OVERALL, IT'S A VERY AESTHETICALLY PLEASING SLIME. THIS IS ONE EXAMPLE OF A KAWAII SLIME, BUT THE POSSIBILITIES ARE ENDLESS!

WHAT YOU'LL NEED

Slime-Making Equipment

Large bowl
Mixing tool (spoon, spatula, or stir stick)
Measuring cups and spoons
An airtight container
Piping bag and tip
Footed glass or plastic cup for slime

Ingredients

One batch of Basic White Glue Slime (see page 12)
1 cup (235 ml) shaving cream

Optional

Color additive (see page 24)
1 to 3 drops of fragrance oil
Acrylic charms
Fake pearls
Glitter

Chocolate & Strawberry Cupcake Slime

Yield: Approximately 16 ounces (475 ml)

A

1 Make a batch of Basic White Glue Slime that's slightly sticky and stretchy (see page 12 for recipe). Optionally, add color and/or fragrance oil. To make the sample slime, I used pink food coloring. Place this slime in a glass or cup (A).

B

2 Fill the piping bag with the shaving cream (B).

3 Pipe a swirl of shaving cream on the top of the slime (C). Decorate with acrylic charms, fake pearls, glitter, or any other fun slime additive.

4 Store the slime in an airtight container so it doesn't dry out. This slime will deflate due to the shaving cream, but you should be able to add more and it will fluff up again. After a while, though, the slime will liquefy and you won't be able to reactivate it.

C

VARIATIONS

Cotton Candy Kawaii Slime = Basic White Glue Slime + pastel pink and blue swirled coloring + shaving cream + cotton candy fragrance oil

Easter Egg Kawaii Slime = Basic White Glue Slime + pastel green coloring + shaving cream + Easter egg charms

Cupcake Kawaii Slime = Basic White Glue Slime + pastel yellow coloring + shaving cream + buttercream fragrance oil

Butterfly Kawaii Slime = Basic White Glue Slime + pastel blue coloring + shaving cream + butterfly wing charms + green glitter

Fresh Peach Kawaii Slime = Basic White Glue Slime + pastel peach coloring + shaving cream + peach charms + green pompoms

5

Hybrid Slimes

ALL THE SLIMES IN THIS CHAPTER ARE A COMBINATION OF TWO DIFFERENT TYPES THAT CREATES AN AMAZING AND UNIQUE TEXTURE. THESE ARE SOME OF MY FAVORITE HYBRID SLIMES, BUT I ENCOURAGE YOU TO EXPERIMENT AND MAKE YOUR OWN HYBRID SLIMES USING DIFFERENT TEXTURES AND RATIOS OF SLIME!

CRUNCHY HYBRID SLIMES

THESE SUPERCRUNCHY SLIMES MAKE GREAT SOUNDS WHEN YOU SQUEEZE THEM!

CHARM SLIME + FISHBOWL SLIME

For this hybrid, I recommend combining equal amounts of clear Charm Slime (recipe on page 32) and Fishbowl Slime (Basic Clear Glue Slime + fishbowl beads).

This slime isn't the stretchiest, but it makes the coolest sounds when all the charms and fishbowl beads crunch together. The hybrid slime isn't crunchier than each individual slime, but it looks so much cooler! The clear fishbowl beads really show off the charms in the slime and makes them pop out more, visually.

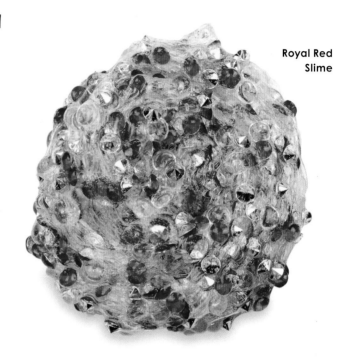

Royal Red Slime

FIZZ SLIME + SLUSHIE SLIME

For this slime, I recommend combining equal amounts of clear glue Fizz Slime (recipe on page 72) and Slushie Slime (Basic Clear Glue Slime + slushie beads).

Fizz Slime and Slushie Slime combine well as they're both crunchy, clear glue–based slimes. Alone, each has a nice texture, and when you combine them, you get the best of both! The resulting slime is very crunchy and makes amazing bubble pop sounds. This slime is perfect for you if you love crunchy slimes.

If you combine them and find that the plastic fake snow or the slushie beads are falling out, add a small dollop of Basic Clear Glue Slime. This should help keep everything together!

On the other hand, if the slime doesn't feel crunchy enough, I would recommend adding more fake snow for the most crunchy slime!

Orange Sherbert Slime

FIZZ SLIME + INFLATING SLIME

I recommend creating a mixture with a ratio of 75% Inflating Slime (recipe on page 40) and 25% clear glue Fizz Slime (recipe on page 72).

You might think this is an odd combination, seeing as Fizz Slime is a crunchy slime and Inflating Slime is smooth and fluffy. But, when you combine these slimes, it creates a slime that makes amazing bubble pops! The fake snow helps incorporate air into the Inflating Slime, changing the texture and creating better bubble pops, which is why this is a Crunchy Hybrid Slime. After you layer this slime, mix it as soon as you can, because the Fizz Slime tends to blend into the Inflating Slime.

Blueberry Batter Slime

FLUFFY HYBRID SLIMES

THESE SLIMES TEND TO BE FLUFFIER AND INFLATE!

CLOUD SLIME + CLEAR SLIME

I recommend adding 75% Basic Clear Glue Slime (recipe on page 12) and 25% Cloud Slime (recipe on page 52).

Mixing Cloud Slime with Basic Clear Glue Slime at this ratio results in a slime that isn't quite a thick Jelly Slime, but isn't a sizzly Icee, either. In the container, before you mix it, the slime looks so cool! The Cloud Slime looks like small mountains frozen in a snow globe. Also, the slime fluffs up a ton when you play with it and may require more activator to be added to it as it becomes quite sticky while mixing.

Splash of Blue Slime

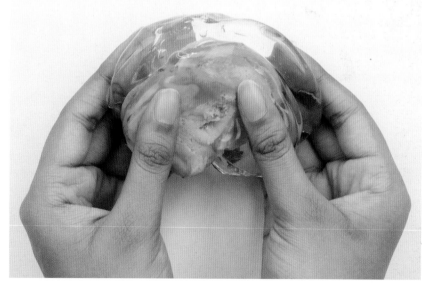

JELLY SLIME + CLEAR GLUE FLOAM

I recommend adding 60% Jelly Slime (recipe on page 46) and 40% Clear Glue Floam (Basic Clear Glue Slime + foam beads).

By adding Clear Glue Floam to Jelly Slime, this adds a crunch to the Jelly Slime. The base of the hybrid is thicker than a plain Clear Glue Slime because there's fake snow in it, and the overall texture when playing with the slime is less sticky than a normal Floam, once activated. One con is that this slime can be a little less stretchy than usual, but overall, it's tons of fun to play with!

Cherry Bomb Slime

INFLATING SLIME + CLEAR GLUE FLOAM

I recommend adding 80% Inflating Slime (see page 40) and 20% Clear Glue Floam (Basic Clear Glue Slime + foam beads).

By adding Clear Glue Floam to Inflating Slime, this creates a sizzlier slime. The foam beads creates a slightly crunchy layer on top of the base of the hybrid slime and is fun to mix in. The texture is smooth with the foam beads almost massaging your hands! This is such a unique slime.

Similar to Fizz Slime + Inflating Slime, after you layer this slime, you want to mix it as soon as you can because the Clear Glue Floam tends to blend into the Inflating Slime.

Sunshine Slime

JELLY CUBE SLIME + GLOSSY SLIME

I recommend adding 75% Basic White Glue Slime (see page 12) and 25% Jelly Cube Slime (page 68).

With this slime, my favorite part of mixing it is crushing all the foam cubes in the slime. This creates a whole new texture with the Glossy Slime, one that inflates a lot. Make sure you don't add too many jelly cubes though, as this can cause the slime to rip. This is why I recommend 25% Jelly Cube Slime and 75% Glossy Slime. Since the majority of the slime is Glossy Slime, the hybrid slime is still stretchy, but the bits of crushed up cubes creates a texture that is really fun to play with!

Oreo Milkshake Slime

**Pop Can Cake Jelly Puff Slime
by Erin Lutterbach Murphy,
The Slimeonade Stand**

INFLATING SLIME + ICEE SLIME

I recommend adding 75% Inflating Slime (recipe on page 40) and 25% Icee Slime (recipe on page 48). By adding this small amount of clear slime and fake snow that's in the Icee Slime, the Inflating Slime has a softer texture that's more sizzly and inflates even more than usual.

Similarly to Fizz Slime + Inflating Slime, after you layer this slime, you want to mix it as soon as you can because the Icee Slime tends to blend into the Inflating Slime.

Summertime Fun Slime

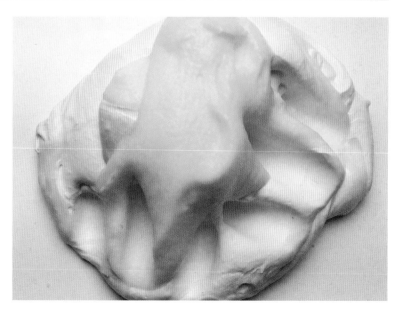

Inflating Slime + Icee Slime—unmixed (right) and swirled (above)—by @wanderlustslime on Instagram

ALL THE FEELS HYBRID SLIMES

EXPERIMENT AND CRAFT YOUR OWN HYBRID SLIME!

It can be tons of fun to mix together a bunch of slimes in a "slime smoothie" to create a unique texture! When doing this, keep in mind the colors you're mixing together because you don't want to end up with a muddled color slime. Lots of slime smoothies turn out gray or brown if you don't think about the colors you're mixing. Also, make sure the scents of the various slimes you're mixing don't clash. Of course, you can mix any slimes together, but these are just some of my handy tips!

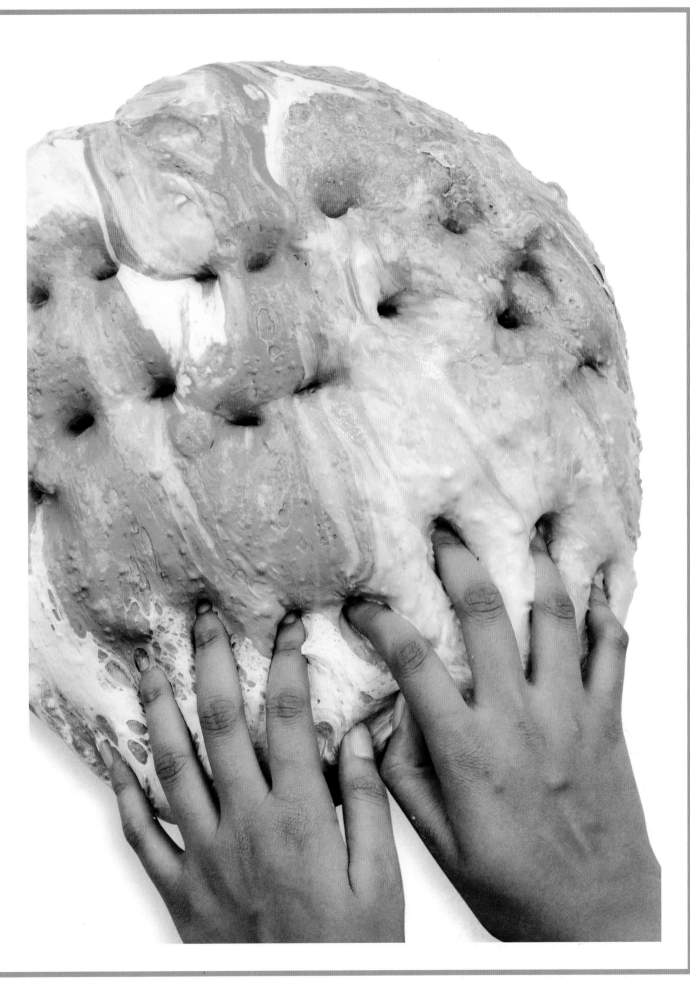

6 Extreme Slime Projects and Ideas

THESE FUN SLIME PROJECTS WILL HAVE YOU PLAYING WITH SLIME IN NEW AND EXCITING WAYS!

BRUSHING ON PIGMENT

MAKING PIGMENTED SLIMES IS ALREADY FUN, BUT BRUSHING ON PIGMENT TAKES PIGMENTED SLIMES TO A NEW LEVEL!

WHAT YOU'LL NEED

8 ounces (235 ml) Basic Clear Glue Slime
   (see page 12)
2 teaspoons (10 ml) pigment
Glitter
Brush

Optional

Charms

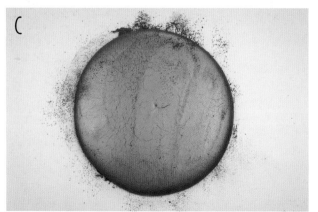

1 Let 8 ounces (235 ml) of slime spread out on a table (A).

2 Sprinkle on the pigment, then use a brush to spread it around the slime in an even layer (B & C).

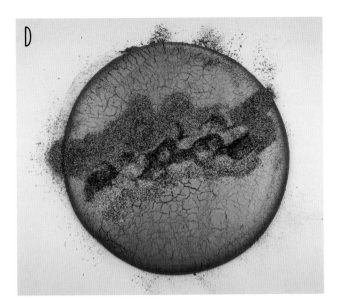

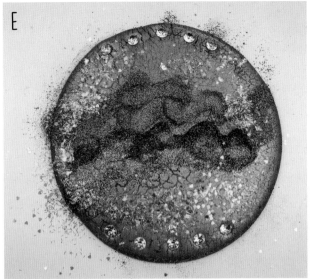

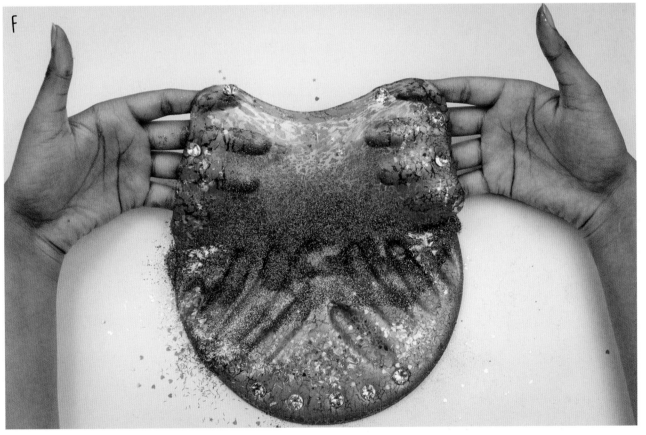

 Sprinkle on the glitter (D).

Optionally, add some charms if you'd like (E)
before mixing (F).

SLIME DESIGNS

THERE ARE SO MANY DIFFERENT WAYS YOU CAN CREATE UNIQUE DESIGNS WITH SLIME. DUE TO THE SLIPPERY NATURE OF SLIME, NO TWO DESIGNS WILL EVER LOOK EXACTLY THE SAME! THIS IS A FUN WAY TO CREATE A "SLIME SMOOTHIE" WHERE YOU MIX A BUNCH OF OLD SLIMES TOGETHER. IF YOU RECORD MAKING YOUR SLIME DESIGN, IT CAN LOOK REALLY INTERESTING TO PLAY THE VIDEO IN REVERSE AFTERWARDS.

1 Lay out slimes in chosen pattern.

2 Use some sort of stick to drag the slime and create a pattern.

3 Mix the slime together!

Here are a couple of examples to get you started on your own journey of making slime designs.

PRETTY PUDDLE

Make Inflating Slime, recipe on page 40. I like using this slime as it doesn't move as quickly as Basic White Glue Slime and you have more time to create the design.

1 Spread out one slime in a large circle. Then, take your next color and spread it out on top of the last layer, but leave a band of the last color visible (A). Repeat as many times as you'd like.

2 Drag straight lines from the middle of the slime out to the edge and from the edge back to the middle (B).

3 Mix!

A
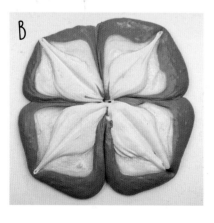

B

Rainbow Puddle Slime by @monalisasliimes

RAINBOW STRIPES

Make Basic Clear Glue Slime, recipe on page 16. Split this up and color each a color of the rainbow using pearl pigment and food coloring.

1 **Lay each slime out in a stripe (A).**

2 **Drag straight lines up and down the slime (B). Move quickly as the slime will spread out!**

3 **Mix (C)!**

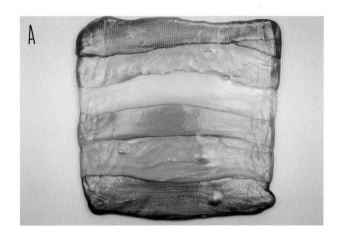

A

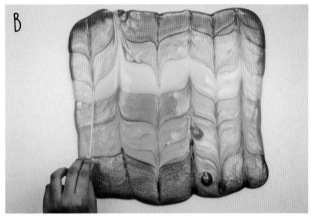

B

C

SLIME AND SQUISHIES

SQUISHIES ARE SHAPES THAT ARE MADE OUT OF FOAM AND PAINTED. THEY ARE SO FUN TO SQUISH FOR STRESS RELIEF! FOR THIS PROJECT, YOU CAN MAKE YOUR OWN SQUISHY AND THEN PUT SLIME ON IT TO CREATE SOME CRUNCHY BUBBLE POPS! UNFORTUNATELY, YOU CAN'T TAKE THE SLIME OUT OF THE SQUISHY AFTER YOU'VE ADDED IT, SO THE SLIME WILL HARDEN IN THE SQUISHY AND YOU'LL HAVE TO THROW IT AWAY AFTERWARDS. I RECOMMEND FILMING THE AWESOME CRUNCHY SOUNDS IT MAKES SO YOU CAN REWATCH THEM AFTER!

FOR THIS PROJECT, I MADE A SIMPLE COOKIE SQUISHY, BUT FEEL FREE TO GET AS CREATIVE AS YOU WANT WITH YOUR SQUISHY MAKING!

WHAT YOU'LL NEED

Circular sponge
Puffy Paint (Light Brown and Dark Brown)
4 ounces (120 ml) of Basic Clear Glue Slime
   (see page 16)

1 Paint the entire sponge in light brown puffy paint **(A)**. Let dry for 12 hours.

B

2 Use dark brown puffy paint to create blobs of "chocolate chips" on your cookie. Let dry for a full 24 hours (B).

3 Take 4 ounces (120 ml) of Basic Clear Glue Slime and place this over the cookie.

4 Let the squishy absorb the slime for 3 to 5 minutes. For the best crunchiness, make sure there isn't any excess slime on the squishy, just what's been absorbed in. Squish!

EXTREME AVALANCHE SLIME

THIS SLIME TAKES BASIC AVALANCHE SLIME TO THE NEXT LEVEL!
THIS HUGE AVALANCHE SLIME IS DEFINITELY AN EXTREME
SLIME THAT'S BOTH GORGEOUS AND INTERESTING.

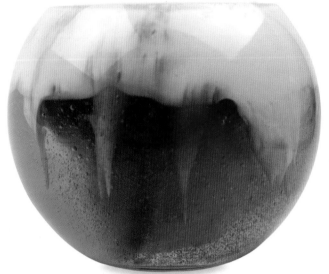

WHAT YOU'LL NEED

One batch of Basic Clear Glue Slime (see page 16)
One batch of Basic White Glue Slime (see page 12)
Large, clear container
Plastic wrap
Back of paintbrush/wooden dowel (something to
    poke thin lines into the slime)

1 Divide your Basic Clear Glue Slime into sections; add color to each. Keep in mind that these colors will be mixing together. For example, pink and blue will mix to a pretty purple, but rainbow colors will mix to gray.

2 Place the Basic Clear Glue slime around the bottom of your container till it is halfway full (A).

3 Optionally, divide your Basic White Glue Slime into sections and color each. I don't color the white glue–based slimes, as I like the focus to be on the vibrant clear glue–based slimes, but it's an option.

4 Place the Basic White Glue Slime on top of the Basic Clear Glue Slime till the container is entirely full (B).

5 Poke the slime 5 to 8 times, depending on how large your container is (C). The larger your container is, the more times you should poke the slime to help it avalanche better. You want to poke the slimes at varying, random heights.

6 Cover the top of the container with plastic wrap and let it sit for 1 to 5 days. Check back every day to see when you're happy with the amount of mixing that has occurred.

7 Mix and play with your slime! The resulting slime makes amazing bubble pops because of the Basic White Glue Slime and Basic Clear Glue Slime mixture.

B

C

DIY CLAY SLIME

THIS SLIME IS SO MUCH FUN TO MIX AND PLAY WITH! IT CREATES A REALLY NICE "SLAY SLIME," MEANING A SLIME WITH A SLIGHT AMOUNT OF CLAY, OR A CLAY SLIME. YOU CAN PUT CLAY ON BOTH THE TOP AND BOTTOM OF THE CONTAINER, BUT IN THIS EXAMPLE I PLACED THE CLAY ON TOP. THIS IS A GREAT WAY TO GET CREATIVE WITH STORING AND GIFTING SLIME. YOU CAN PUT THIS PROJECT TOGETHER, FOR SOMEONE, THEN THEY CAN MIX IT THEMSELVES!

WHAT YOU'LL NEED

Ingredients

7 ounces (210 ml) Basic White Glue Slime
   (see page 12)
1 ounce (28 ml) air-dry clay
8-ounce (235 ml) container

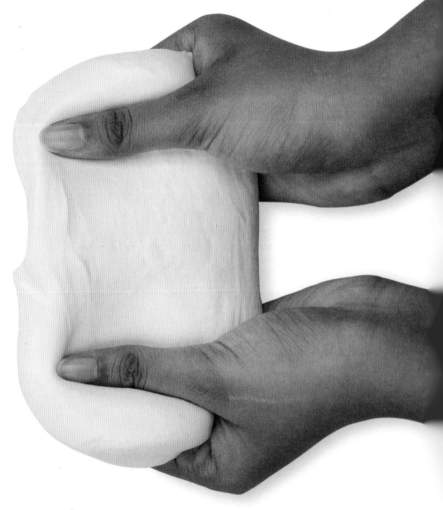

A

1 Take clay and roll it out to the thickness you want. I typically make mine ¼ inch (6 mm) thick. Use the container to press into the clay and cut out a circle. Set aside.

2 Fill an 8 ounce (235 ml) slime container with about 7 ounces (210 ml) of Basic White Glue Slime, or enough so there's just enough space for the clay circle you cut out in step 1 (A).

3 Place the circle of clay on top of the Basic White Glue Slime (B). The clay should rest flush to the top rim of the container. You don't want any space here, as any air in the container will dry out the clay.

4 Press the clay into the slime and mix the two together (C)! If you aren't immediately mixing them, don't wait more than 2 weeks to mix it, as the clay will dry out and the slime will become sticky, making it harder to mix. You can still wait that long and add activator, but that isn't ideal.

B

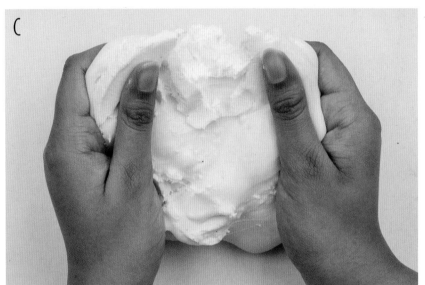

C

DIY Clay Slime by @SomeOtherSlime on Instagram

SLIME PALETTES

THESE SLIME PALETTES ARE SO AESTHETICALLY PLEASING AND A UNIQUE
WAY TO STORE SLIME! THERE ARE SO MANY OPTIONS FOR COLORS AND TEXTURES
OF SLIME THAT YOU CAN STORE IN THESE PALETTES.

WHAT YOU'LL NEED

Container with partitions (the
    partitions should *not* be
    removable; if they are, the
    slime will seep under them)
Slime in any color or texture!

A

All you need to do is place the slimes into the palette! Make sure that each section is filled fully to the top so there's no air to dry out the slime. Also make sure they aren't overfilled, because then the slime will spill over into other sections. Here are some examples.

RAINBOW SLIME PALETTE

Fill the palette with rainbow colors! For the most vibrant colors in Basic White Glue Slime, use acrylic paint to color it. I also included gold, silver, and copper slime to fill up the rest of the palette **(A)**.

B

OMBRE SLIME PALETTE

This palette features various shades of the same color. To do this, add 1 drop of food coloring, then two to the next section, three to the next one, and four to the last. Repeat this for all the colors. I also added some clear slime with glitter and a pearlescent slime because they matched and looked awesome together **(B)**!

C

PINK, GREEN, BLUE AND PURPLE PALETTE

You can also create a color-specific palette. So, you create a bunch of different types of slime with the same color. I made Pompom Slime, Metallic Slime, and Glitter Slime in bright pink, green, blue, and purple **(C)**.

FLOAM EXPLOSIONS

THIS SLIME IS SO CRUNCHY AND PRETTY! THE GLITTER IS GORGEOUS, AND THE FOAM BEADS CREATE THE BEST TEXTURE. THIS IS SUCH A FUN WAY TO PLAY WITH SLIME.

WHAT YOU'LL NEED

8 ounces (235 ml) of Basic Clear Glue Slime (see page 16)

Foam beads (same amount as slime by volume)

Glitter (approximately 2 to 4 ounces or 60 to 120 ml)

1 Make a Clear Glue Floam. Mix in equal amounts of foam beads to sticky Basic Clear Glue Slime (A).

2 Lay out the Floam flat in a circle. Pour glitter into the middle of the slime (B).

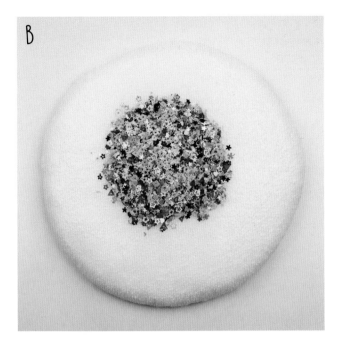

B

C

3 Gather all the edges of the slime together. Roll the slime into a ball so that the center of the ball is filled with glitter (C). You can also press the sides flat to create a Floam cube.

4 Stretch the slime apart (D). Mix together the glitter and foam beads.

5 Store the slime in an airtight container so it doesn't dry out.

D

SLIME SCULPTURES

THERE ARE SO MANY DIFFERENT WAYS TO MAKE SLIME SCULPTURES! YOU CAN USE ANY BRAND OF CLAY AND WHITE OR CLEAR GLUE. IF YOU USE MORE CLAY THAN SLIME, THE RESULT WILL BE A BUTTER SLIME, AND IF YOU USE LESS CLAY THAN SLIME, THE RESULT WILL BE A CLAY SLIME.

THE PROJECTS BELOW AND ON THE OPPOSITE PAGE WERE CONTRIBUTED BY SOME OF MY FRIENDS IN THE SLIME COMMUNITY. TO CREATE MY PROJECTS (SEE PAGES 106–107), I STARTED BY MOLDING WHITE CLAY, THEN USED CHALK PASTELS TO ADD COLOR.

WHAT YOU'LL NEED

Clay: You can use any clay, but I find that Crayola Model Magic is easiest to sculpt and doesn't dry out while sculpting white glue– or clear glue–based slimes.

Optional

Chalk pastels
Blade
Brush
Glitter

Clay Slime Gift Sculpture by @SomeOtherSlime on Instagram

HOW TO USE CHALK PASTELS

1. Shave off some of the chalk pastel using a blade onto a piece of paper or clean surface.
2. Brush it onto the clay using a paintbrush.

Chocolate-Dipped Strawberry Sculpture by @wanderlustslime on Instagram

Spaghetti Slime Sculpture by @monalisasliimes

SUNFLOWER FIELD

1. Hand sculpt flowers, or use a mold.

2. Roll out the stem and leaves from green clay.

3. Place all the clay onto Basic Clear Glue Slime.

4. Pour on blue glitter, representing the sky.

5. Mix!

BREAD BUN

1. Roll out a ball of white clay.

2. Brush on a mixture of yellow, orange, and brown chalk pastel.

3. To create the crackled pattern on top, make sure that you use a drier clay and pull it apart slightly.

4. Place the fake bread onto Basic Clear Glue Slime.

5. Mix!

CHOCOLATE AND ROSES

1 Roll out a snake of pink clay. Swirl it together.

2 Hand sculpt the chocolate or use a mold.

3 Place the fake rose and chocolate onto Basic Clear Glue Slime.

4 Mix!

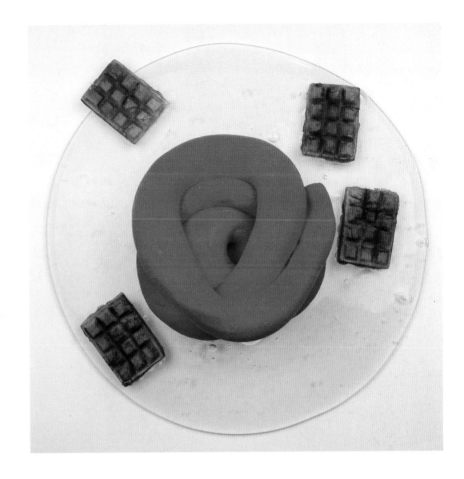

ERUPTING VOLCANO

1 Sculpt a brown volcano from clay.

2 Place the fake volcano onto Basic Clear Glue Slime.

3 Drip red, gold, and copper slime down the sides and on top.

4 Mix!

CONTRIBUTING SLIMERS

Chelsey P., H0nestslimereviews and H0nestlyslime

@monalisasliimes

Marjorie Lounds, @RainbowPlayMaker

Erin Lutterbach Murphy, The Slimeonade Stand

@SomeOtherSlime on Instagram

@wanderlustslime on Instagram

RESOURCES

I shop at craft stores, dollar stores, arts and crafts sections of general stores, and online for my supplies. Included below are some of my preferred brands.

BASIC SLIME INGREDIENTS

Elmer's Glue
www.elmers.com

Foaming Hand Soap
www.bathandbodyworks.com/c/hand-soaps/
foaming-hand-soap

Renu Fresh Multipurpose Contact Lens Cleaning Solution
www.renu.com

Baking Soda
www.armandhammer.com/baking-soda

Tide Free and Gentle Liquid Laundry Detergent
www.tide.com/en-us/shop/type/liquid/
tide-free-and-gentle-liquid

Sta-Flo Liquid Starch
www.purex.com/products/laundry-enhancers/
sta-flo-liquid-starch

ADD-INS

Acrylic Paint
www.plaidonline.com/brands/folkart

Color-Changing Pigment
www.solarcolordust.com

Crayola Model Magic
www.crayola.ca/things-to-do/how-to-landing/
model-magic.aspx

Daiso Soft Clay
www.daisojapan.com

Fragrance Oil
www.justscent.com

Lotion
www.bathandbodyworks.com/c/body-care/
body-lotion

Nendo Soft Clay
www.nendosoftclay.com/shop

Pearl Ex Pigments
www.jacquardproducts.com/pearl-ex

Polymer Clay
www.sculpey.com/sculpey-iii/14-sculpey-iii

SnoWonder
www.snowonder.com

FIND THESE PRODUCTS AT YOUR LOCAL RETAILER OR ONLINE

- Ashland Clear Decorative Filler
- Baby Bumpers: M2cbridge Multifunctional Edge and Corner Guard Coverage Baby Safety Bumpers
- Chalk Pastels: Artist's Loft
- FloraCraft Winter Snow
- Foam Sheets: Darice Foamies Foam Sheets
- Food Coloring: McCormick's
- Puffy Paint: Scribbles Dimensional Fabric Paint
- Recollections Glitter

CONNECT WITH ME!

- Instagram: @craftyslimecreator
- YouTube: Alyssa J
- Etsy: craftedbyalyssaj

ACKNOWLEDGMENTS

It truly takes a village to create a book. I have so many people to thank for making my book a possibility. I am so grateful to

- Amanda DeMatos and Mandy Persaud, my cheerleaders.

- My teachers and mentors, especially Ms. Phillips for teaching me to follow my passions.

- The awesome slimers who contributed to this book. You are all such creative, inspiring people.
  Sasha (@SomeOtherSlime)
  Mia (@wanderlustslime
  Marjorie (@rainbowplaymaker
  Nikki (@monalisasliimes)
  Erin (@slimeonadestand)
  Chelsey (@H0nestslimereviews)

- All my friends and supporters online who I may never meet face to face, but whose support is invaluable—Karen from @kreative_rainbow and Penny from @froggsspittslimes.

- The entire Quarto team: Hannah Moushabeck, Lydia Anderson, John Gettings, Marissa Giambrone, Barb States, Kristine Anderson, Ken Fund, Regina Grenier, Mary Ann Hall, and Winnie Prentiss. You never cease to amaze me in your attention to detail and dedication to my book.

- Sam Welbourn, the best photographer I could ask for.

- Joy Aquilino, my amazing editor. Thank you so much for championing my books and making them a reality.

- My family—Dad, Nanee, Nana, and Cierra. Thank you for your constant love and support.

- And, thank you to my biggest supporter who has helped me through all the highs and lows, Mom.

ABOUT THE AUTHOR

Alyssa Jagan is Alyssa of @CraftySlimeCreator, which has been named one of the best slime accounts on Instagram. She is also the author of *Ultimate Slime*, which has been published in eleven languages, and the coauthor of *Study with Me*, on how to use bullet journaling and time-management techniques for successful studying. An 18–year–old college student and Toronto native, the popular Instagram slimer posts videos to her Instagram accounts every day and cohosts the Slimey Sundays live podcast on Instagram with Erin Murphy of @slimeonadestand and sells products on her Etsy site, craftedbyalyssaj. In addition, Alyssa has been profiled by a variety of media outlets, including *The New York Times*, Associated Press, *The Globe and Mail*, and the BBC.

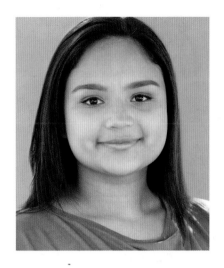

INDEX

A

Acrylic charms, about, 27
Activator types, 10
Air-dry clay types, about, 38
Aloha Slime, 59
Autumn Slime, 59

B

Basic Clear Glue Slime, 16–17
Basic White Glue Slime, 12–13
Beads, about, 26
Birthday Cake Batter, 41
Black Tie Slime, 69
Blueberry Bread Slime, 43
Bonfire Slime, 41
Bread Bun Sculptures, 106
Bread Slime, 42–43
Brownie Batter Slime, 61
Brushing on Pigment
    Project, 90–91
Bumblebee Fizz Slime, 73
Butterfly Kawaii Slime, 75
Butterscotch Cloud
    Creme, 51

C

Candy Apple Slime, 69
Candy Corn Slime, 63
Candy Crush Slime, 59
Candy Land Slime, 65
Candy Store Slime, 67
Caribbean Salsa Col-
    or-Changing Slime, 57
Cereal Slime, 62–63
Chalk Pastels, using, 105
Champagne Slime, 71
Charm Slime + Fishbowl
    Slime, 78
Charms
    acrylic, 27
    making, 31
Cheerios Slime, 63
Cherry Bomb, 41
Cherry Jolly Rancher Slime, 65
Chewy Bubblegum Slime, 41
Chocolate and Roses
    Sculptures, 107
Chocolate Chip Cookie
    Dough Slime, 61
Chocolate Chunk Slime,
    60–61
Cinnamon Roll Slime, 63
Cleaning up, 20
Cloud Cream, 50–51
Cloud Slime, 52–53
Cloud Slime + Clear Slime, 81
Cola Slime, 69
Color-Changing Slime, 56–57
Colorants, about, 24
Cornbread Slime, 43
Cotton Candy Kawaii
    Slime, 75
Crayola Model Magic,
    about, 38

Crunchy Charms Slime,
    66–67
Cupcake Kawaii Slime, 75

D

Daiso clay, about, 38
Diva Slime, 71
DIY Clay Slime, 98–99

E

Easter Egg Kawaii Slime, 75
Elmer's glue types, 10
Erupting Volcano
    Sculptures, 107
Eucalyptus Icee Slime, 49
Extra-Thick White Glue Slime,
    14–15
Extreme Avalanche Slime,
    96–97

F

Fake cereal, 29–30
Fake chocolate chunks,
    29–30
Fake snows, 44–45
Fake sprinkles, 30
Fishbowl beads, about, 26
Fizz Slime, 72–73
Fizz Slime + Inflating Slime, 80
Fizz Slime + Slushie Slime, 79
Flamingo Feathers Cloud
    Slime, 53
Flower Power Slime, 67
Foam beads, about, 26
Foam chocolate chunks, 35
Foam cubes, 33
Foam Explosions, 102–103
Foam shapes, 34
Foam Shapes Slime, 64–65
Fragrance oils, 25
French Baguette Slime, 43
Fresh Green Apple Icee
    Slime, 49
Fresh Peach Kawaii Slime, 75
Frozen Slime, 71
Fruit Loops Slime, 63
Fruit Salad Slime, 67
Fuzzy Blanket, 41

G

Galaxy Goo Slime, 69
Galaxy Pop Slime, 73
Gingerbread Slime, 43
Glitter, about, 25
Glycerin, about, 11
Green Jolly Rancher
    Slime, 59
Groovy Color-Changing
    Slime, 57

H

Hand soap, foaming, 11
Hard Slime, fixing, 20
Head in the Clouds Slime, 53

I

Icee Slime, 48–49
Inflating Slime, 40–41
Inflating Slime + Clear Glue
    Slime, 83
Inflating Slime + Icee Slime,
    85

J

Java Chip Slime, 60–61
Jelly Bean Slime, 59
Jelly Cube Slime, 68–69
Jelly Cube Slime + Glossy
    Slime, 84
Jelly Slime, 46–47
Jelly Slime + Clear Glue
    Slime, 82
Jellyfish Jelly Slime, 47

K

Kawaii Slime, 74–75

L

Lemon Slime, 69
Lotions, about, 11
Lucky Penny Jelly Slime, 47

M

Mango Twist Icee Slime, 49
Mermaid Cloud Creme, 51
Metallic Foil Slime, 70–71
Metallic foil, about, 28
Midnight Jelly Slime, 47
Mystical Jelly Slime, 47

N

Nendo soft clay, about, 38

O

Ocean Cruise Slime, 67
Ocean Waves Icee Slime, 49
Orange Soda Fizz Slime, 73
Oreo Color-Changing
    Slime, 57

P

Peppermint Slime, 61
Pigments, about, 24
Pineapple Party Cloud
    Creme, 51
Pink Candy Color-Changing
    Slime, 57
Planet Earth Color-Changing
    Slime, 57
Plastic beads, about, 26
Polyvinyl acetate (PVA)
    glue, 10
Pompom Slime, 58–59
Pompoms, about, 27
Pop of Pastel Slime, 63
Pretty Puddle Slime
    Design, 92

R

Rainbow Stripes, 93
Red Velvet Slime, 61
Rose Water Jelly Slime, 47

S

Safety tips, 7
Sea Foam Slime, 73
Slime and Squishies, 94–95
Slime Palettes, 100–101
Slime Sculptures, 104–107
Slushie beads, about, 26
S'mores Slime, 61
Snowed-in Slime, 65
SnoWonder/Instant Snow,
    44, 45
Soap, about, 11
Soothing Lavender Cloud
    Slime, 53
Spiced Cinnamon Cloud
    Creme, 51
Squishies, 94
Starry Night Slime, 65
Sticky Slime, fixing, 19
Stiff Slime, fixing, 20
Storing slime, 21
Strawberry Bread Slime, 43
Strawberry Lemonade Icee
    Slime, 49
Sugar beads, about, 26
Summer Daze Cloud
    Slime, 53
Summer Sun Slime, 71
Sunflower Field Sculptures, 106
Super Absorbent Polymers
    (SAP) snow, 44, 45

T

Tough top layer, fixing, 18

V

Valentine's Slime, 65
Vanilla Bean Cloud
    Creme, 51
Vanilla Rose Cloud Slime, 53
Victorious Slime, 71

W

Wicked Fizz Slime, 73
Willy Wonka Slime, 67